Student Solutions M

Statistical Methods for Engineers

Geoffrey Vining

Virginia Polytechnic and State University

Scott Kowalski

Minitab, Inc.

Prepared by

Scott Kowalski

Minitab, Inc.

BROOKS/COLE
CENGAGE Learning

Australia • Brazil • Japan • Korea • Mexico • Singapore • Spain • United Kingdom • United States

ISBN-13: 978-0-538-73880-4
ISBN-10: 0-538-73880-4

Brooks/Cole
20 Channel Center Street
Boston, MA 02210
USA

Cengage Learning is a leading provider of customized learning solutions with office locations around the globe, including Singapore, the United Kingdom, Australia, Mexico, Brazil, and Japan. Locate your local office at: **www.cengage.com/global**

Cengage Learning products are represented in Canada by Nelson Education, Ltd.

To learn more about Brooks/Cole, visit **www.cengage.com/brookscole**

Purchase any of our products at your local college store or at our preferred online store **www.CengageBrain.com**

Printed in the United States of America
1 2 3 4 5 6 7 14 13 12 11 10

TABLE OF CONTENTS

Chapter 1   *Overture: Engineering Method and Data Collection* 1

Chapter 2   *Data Displays* 5

Chapter 3   *Modeling Random Behavior* 29

Chapter 4   *Estimation and Testing* 49

Chapter 5   *Control Charts and Statistical Process Control* 83

Chapter 6   *Linear Regression Analysis* 123

Chapter 7   *Introduction to 2^k Factorial Experiments* 149

Chapter 8   *Introduction to Response Surface Methodology* 163

CHAPTER 1

OVERTURE: ENGINEERING METHOD AND DATA COLLECTION

1.1 **a** coating thickness

    b viscosity (continuous)

    c low and high

    d with only one factor, this is the same as part c

    e the experimental unit and the observational unit are an individual part

    f randomly select panels from the two levels of viscosity

1.3 **a** plate thickness after 200 charge-discharge cycles

    b porosity (continuous)

    c low and high

    d with only one factor, this is the same as part c

    e the experimental unit and the observational unit are a plate

    f for each ED bath, randomly select 5 plates from both low and high porosity

1.5 **a** rating

    b method (categorical)

    c motor and research

    d with only one factor, this is the same as part c

    e the experimental unit and the observational unit are a blend

    f randomly select a blend of a certain octane, divide the blend in two, and randomly assign each half to a method

1

1.7   **a**   maximum output (in fluid ounces) per hour

b   brand of humidifier (categorical)

c   A and B

d   only one factor, this is the same as part c

e   the experimental unit and the observational unit are a humidifier

f   assign the 16 humidifiers to the chamber in random order

1.9   **a**   amount of chlorine

b   vendor (categorical)

c   A, B, C, D, E

d   only one factor, this is the same as part c

e   the experimental unit and the observational unit are a sponge

f   test the 20 sponges in random order

1.11 **a**   pressure to separate the cap from the bottle

b   injection speed, mold temperature and cooling time (all continuous)

c   injection speed (40 & 75), mold temperature (25 & 45), cooling time (10 & 25)

d

Speed	Temperature	Time
40	25	10
75	25	10
40	45	10
75	45	10
40	25	25
75	25	25
40	45	25
75	45	25

e (1)   the experimental unit and the observational unit are a cavity

(2)   for each of the 4 shots, randomly assign a treatment to a cavity

f (1)   the experimental unit is a shot and the observational units are the cavities

(2)   assign the treatments to the shots in random order

CHAPTER 1

OVERTURE: ENGINEERING METHOD AND DATA COLLECTION

1.1 **a** coating thickness

   b viscosity (continuous)

   c low and high

   d with only one factor, this is the same as part c

   e the experimental unit and the observational unit are an individual part

   f randomly select panels from the two levels of viscosity

1.3 **a** plate thickness after 200 charge-discharge cycles

   b porosity (continuous)

   c low and high

   d with only one factor, this is the same as part c

   e the experimental unit and the observational unit are a plate

   f for each ED bath, randomly select 5 plates from both low and high porosity

1.5 **a** rating

   b method (categorical)

   c motor and research

   d with only one factor, this is the same as part c

   e the experimental unit and the observational unit are a blend

   f randomly select a blend of a certain octane, divide the blend in two, and randomly assign each half to a method

1

1.7   **a**   maximum output (in fluid ounces) per hour

    b   brand of humidifier (categorical)

    c   A and B

    d   only one factor, this is the same as part c

    e   the experimental unit and the observational unit are a humidifier

    f   assign the 16 humidifiers to the chamber in random order

1.9   **a**   amount of chlorine

    b   vendor (categorical)

    c   A, B, C, D, E

    d   only one factor, this is the same as part c

    e   the experimental unit and the observational unit are a sponge

    f   test the 20 sponges in random order

1.11   **a**   pressure to separate the cap from the bottle

    b   injection speed, mold temperature and cooling time (all continuous)

    c   injection speed (40 & 75), mold temperature (25 & 45), cooling time (10 & 25)

    d

Speed	Temperature	Time
40	25	10
75	25	10
40	45	10
75	45	10
40	25	25
75	25	25
40	45	25
75	45	25

    e (1)   the experimental unit and the observational unit are a cavity

       (2)   for each of the 4 shots, randomly assign a treatment to a cavity

    f (1)   the experimental unit is a shot and the observational units are the cavities

       (2)   assign the treatments to the shots in random order

1.13 a   vertical component of a dynamometric reading

b   angle of edge level (continuous) and type of cut (categorical)

c   angle of edge level (15 & 30), type of cut (continuous & interrupted)

d

Angle	Cut
15	continuous
30	continuous
15	interrupted
30	interrupted

e   the experimental unit and observational unit are a piece of metal

f   for each piece of metal, carry out the four cuts in random order

1.15 a   texture of a cake

b   amount of flour, amount of egg powder, amount of oil, temperature of oven (all continuous)

c   amount of flour (low & high), amount of egg powder (low & high), amount of oil (low & high), temperature of oven (375 & 400)

d

Flour	Egg Powder	Oil	Temperature
low	low	low	375
high	low	low	375
low	high	low	375
high	high	low	375
low	low	high	375
high	low	high	375
low	high	high	375
high	high	high	375
low	low	low	400
high	low	low	400
low	high	low	400
high	high	low	400
low	low	high	400
high	low	high	400
low	high	high	400
high	high	high	400

e   No

f   the experimental unit and observational unit are the oven

g   the experimental unit and observational unit are a cake

TABLE OF CONTENTS

Chapter 1   *Overture: Engineering Method and Data Collection* 1

Chapter 2   *Data Displays* 5

Chapter 3   *Modeling Random Behavior* 29

Chapter 4   *Estimation and Testing* 49

Chapter 5   *Control Charts and Statistical Process Control* 83

Chapter 6   *Linear Regression Analysis* 123

Chapter 7   *Introduction to 2^k Factorial Experiments* 149

Chapter 8   *Introduction to Response Surface Methodology* 163

CHAPTER 2

DATA DISPLAYS

2.1

Stem Leaves	No.	Depth
.37 6	1	1
.37 777777	6	7
.37 888888888	9	16
.37 99999999999999	14	
.38 0000	4	5
.38 1	1	1

n = 35
center (median) = .379; range from .376 to .381
single peak with no outliers, skewed left.

2.3

Stem Leaves	No.	Depth
.6* 3	1	1
.6● 67788899	8	9
.7* 1122244	7	16
.7● 56666777899	11	
.8* 00223	5	8
.8● 667	3	3

n = 35
center (median) = .76; range from .63 to .87
single peak with no outliers, approximately symmetric.

2.5

Stem Leaves	No.	Depth
342.f 4	1	1
342.s	0	1
342.●	0	1
343.* 011	3	4
343.t 33333	5	
343.f 445	3	5
343.s 7	1	2
343.● 8	1	1

n = 14
center around 343.3; range from 342.4 to 343.8
single peak with one outlier (342.4), skewed right.

2.7

Stem	Leaves	No.	Depth
.8*	01	2	2
.8t		0	2
.8f		0	2
.8s		0	2
.8●	9	1	3
.9*		0	3
.9t	23	2	5
.9f	4455	4	9
.9s	66677	5	
.9●	89999999999	11	11

n = 25
center around .97; range from .80 to .99
single peak with two possible outlier (0.80 and 0.81), skewed left.

2.9

Stem	Leaves	No.	Depth
20.f	0	1	1
20.s	222	3	4
20.●		0	4
20.*	677777	6	10
20.t	89	2	
30.f	01	2	10
30.s	2333333	7	8
30.●	4	1	1

Data has decimal value truncated
n = 22
center (median) = 29.025; range from 20 to 34
two peaks with no outliers.

2.11

Stem	Leaves	No.	Depth
0	111223333344455566667889	21	21
1	00145679	8	29
2	889	3	32
3	0236	4	36
4	024	3	39
5	0124568	7	
6	345	3	38
7	01589	5	35
8	4689	4	30
9	0127	4	26
10	034456	6	22
11	024578	6	16
12	0124568999	10	10

Data has decimal value truncated
$n = 84$
center (median) = 53; range from 1 to 129
two peaks with no outliers, electronics tend to fail either right at the beginning or they last a fairly long time which is where the two peaks are located.

2.13

Stem	Leaves	No.	Depth
0.5	5	1	1
0.6		0	1
0.7	47	2	3
0.8	14	2	5
0.9	3	1	6
1.0	4	1	7
1.1	13	2	9
1.2	45789	5	14
1.3	069	3	17
1.4	28899	5	22
1.5	0012345589	10	
1.6	0111122346667889	16	31
1.7	0036678	7	15
1.8	12449	5	8
1.9		0	3
2.0	01	2	3
2.1		0	1
2.2	4	1	1

$n = 63$
center (median) = 1.59; range from .55 to 2.24
skewed left with outliers at both ends.

2.15

	No Cover				*Cover*	
Depth	No.	Leaves	Stem	Leaves	No.	Depth
0	0		.35	9	1	1
0	0		.36	2	1	2
2	2	64	.37	67	2	4
2	0		.38	3346	3	8
	6	653220	.39	7	1	
5	2	95	.40	1459	4	7
6	3	842	.41	1	1	3
3	1	1	.42	45	2	2
2	2	55	.43		0	0

$n_1 = n_2 = 16$
No Cover centered .4005; ranges from .374 to .435
Cover centered .3915; ranges from .359 to .425
Both No Cover and Cover are reasonably symmetric

2.17

	23 seconds				*25 seconds*	
Depth	No.	Leaves	Stem	Leaves	No.	Depth
			7.*	1	1	1
3	3	322	7.t	33	2	3
9	6	555554	7.f	445	3	6
	11	76666666666	7.s	6677	4	10
4	4	8888	7.●	888999999	9	
			8.*	011	3	5
			8.t	22	2	2

$n_1 = n_2 = 24$
Free height at 23 seconds centered 7.6; ranges from 7.2 to 7.8
Free height at 25 seconds centered 7.8; ranges from 7.1 to 8.2
Free height at 23 seconds has a single peak
Free height at 25 seconds has a single peak, slightly skewed left
25 seconds tends to be closer to the target of 8 but there is more variation

2.19

	Low Viscosity				*High Viscosity*	
Depth	No.	Leaves	Stem	Leaves	No.	Depth
3	3	888	0●	7789	4	4
	7	4432100	1*	44444	5	
6	6	877665	1●	555	3	7
			2*	0113	4	4

$n_1 = n_2 = 16$
Data split between ones and tenths and truncated at the tenths place.
Low viscosity centered around 1.4; ranges from 0.83 to 1.83
High viscosity centered around 1.4; ranges from 0.74 to 2.36
Low viscosity is fairly symmetric with a single peak and no outliers.
High viscosity is evenly distributed with no outliers.

Low viscosity has lower variability than high viscosity. For both high and low viscosity, the thickness values are far from the target of .8mm.

2.21

			Reflux Rate of 70		*Reflux Rate of 80*		
Depth	No.	Leaves	Stem	Leaves		No.	Depth
2	2	98	7.●	69		2	2
2	0		8.*	0111		4	6
3	1	7	8.●	9		1	7
9	6	432221	9.*	00233444		8	
	16	9999997776665555	9.●	5556777799		10	10

$n_1 = n_2 = 25$

Reflux Rate of 70 is centered around .95; ranges from .78 to .99
Reflux Rate of 80 is centered around .94; ranges from .76 to .99
Reflux Rate of 70 has a single peak and is skewed left with two outliers (.78 and .79)
Reflux Rate of 80 has two peaks and is skewed left
25 seconds tends to be closer to the target of 8 but there is more variation

Reflux rate of 70 appears to have smaller variation.

2.23 $n = 50$, $l_m = (50 + 1)/2 = 25.5$, $l_q = (50 + 2)/4 = 13$

median $= (y_{(25)} + y_{(26)})/2 = (8.0 + 8.0)/2 = 8.0$
$Q_1 = y_{(13)} = 7.8$ $Q_3 = y_{(38)} = 8.1$

step $= 1.5(Q_3 - Q_1) = 1.5(.3) = .45$
UIF $= 8.1 + .45 = 8.55$ UOF $= 8.1 + 2(.45) = 9.0$
LIF $= 7.8 - .45 = 7.35$ LOF $= 7.8 - 2(.45) = 6.9$

```
            _____
  |-----|_____|__|--------|

|--|--|--|--|--|--|--|--|--|--|
75 76 77 78 79 80 81 82 83 84 85
```

centered around 8.0; bulk of data between 7.8 and 8.1; data ranges from 7.6 to 8.4
approximately symmetric, with no outliers

2.25 $n = 100$, $l_m = (100 + 1)/2 = 50.5$, $l_q = (100 + 2)/4 = 25.5$

median $= (y_{(50)} + y_{(51)})/2 = (2.7 + 2.7)/2 = 2.7$
$Q_1 = (y_{(25)} + y_{(26)})/2 = (1.8 + 1.8)/2 = 1.8$
$Q_3 = (y_{(75)} + y_{(76)})/2 = (3.2 + 3.2)/2 = 3.2$

step $= 1.5(Q_3 - Q_1) = 1.5(1.4) = 2.1$
UIF $= 3.2 + 2.1 = 5.3$ UOF $= 3.2 + 2(2.1) = 7.4$
LIF $= 1.8 - 2.1 = -0.3$ LOF $= 1.8 - 2(2.1) = -2.4$

```
       |-------------|_____|____|----------------|       0


  _____
    |     |     |     |     |     |     |     |     |     |     |     |     |     |
    0    0.5   1.0   1.5   2.0   2.5   3.0   3.5   4.0   4.5   5.0   5.5   6.0
```

centered around 2.7; bulk of data between 1.8 and 3.2; data ranges from 0.4 to 5.6
approximately symmetric with one mild outlier (5.6)

2.27 $n = 25$, $l_m = (25 + 1)/2 = 13$, $l_q = (25 + 3)/4 = 7$

median $= y_{(13)} = 18.1$ $Q_1 = y_{(7)} = 13.8$ $Q_3 = y_{(25 + 1 - 7)} = y_{(19)} = 21.5$

step $= 1.5(Q_3 - Q_1) = 1.5(7.7) = 11.55$
UIF $= 21.5 + 11.55 = 33.05$ UOF $= 21.5 + 2(22.8) = 44.6$
LIF $= 13.8 - 11.55 = 2.25$ LOF $= 13.8 - 2(11.55) = -9.3$

```
       |-----|_|__|---|     0    0         *                    *


  _____
    |         |         |         |         |         |         |         |         |
    0        10        20        30        40        50        60        70        80
```

centered around 18; bulk of data between 13.8 and 21.5; data ranges from 8.0 to 79.2
skewed right with two mild outliers (35.1, 40.3) and two extreme outliers (52.3, 79.2).

2.29 $n = 26$, $l_m = (14 + 1)/2 = 13.5$, $l_q = (26 + 2)/4 = 7$

median $= (y_{(13)} + y_{(14)})/2 = (119 + 119)/2 = 119$

$Q_1 = y_{(7)} = 87$ $\qquad$ $Q_3 = y_{(26 + 1 - 7)} = y_{(20)} = 182$

step $= 1.5(Q_3 - Q_1) = 1.5(95) = 142.5$

UIF $= 182 + 142.5 = 324.5$ $\quad$ UOF $= 182 + 2(142.5) = 467$

LIF $= 87 - 142.5 = -55.5$ $\quad$ LOF $= 87 - 2(142.5) = -198$

```
           |----|__|_____|----------|                          *
_____
     |     |     |     |     |     |     |     |     |     |     |
     0    50   100   150   200   250   300   350   400   450   500   550
```

entered around 119; bulk of data between 87 and 182; data ranges from 30 to 511. skewed right with one extreme outlier (511).

2.31 $n = 50$ $\quad$ $l_m = (50 + 1)/2 = 25.5$, $l_q = (50 + 2)/4 = 13$

median $= (y_{(25)} + y_{(26)})/2 = (2.0018 + 2.0018)/2 = 2.0018$

$Q_1 = y_{(13)} = 2.0015$ $\quad$ $Q_3 = y_{(50 + 1 - 13)} = y_{(38)} = 2.0021$

step $= 1.5(Q_3 - Q_1) = 1.5(.0006) = .0009$

UIF $= 2.0021 + .0009 = 2.003$, $\quad$ UOF $= 2.0021 + 2(.0009) = 2.0039$

LIF $= 2.0015 - .0009 = 2.0006$, $\quad$ LOF $= 2.0015 - 2(.0009) = 1.9997$

```
           |---------|_____|_____|-------------------------|
_____
   2.0012 2.0014 2.0016 2.0018 2.0020 2.0022 2.0024 2.0026 2.0028
```

center (median) = 2.0018; bulk of data between 2.0015 and 2.0021, range from 2.0013 to 2.0027; skewed right with no outliers

2.33 n = 47 $l_m = (47 + 1)/2 = 24$, $l_q = (47 + 3)/4 = 12.5$

median = $y_{(24)} = 1331$
$Q_1 = (y_{(12)} + y_{(13)})/2 = (850 + 880)/2 = 865$ $Q_3 = (y_{(35)} + y_{(36)})/2 = (1717 + 1737)/2 = 1727$

step = $1.5(Q_3 - Q_1) = 1.5(862) = 1293$
UIF = $1727 + 1293 = 3020$, UOF = $1727 + 2(1293) = 4313$
LIF = $865 - 1293 = -428$, LOF = $865 - 2(1293) = -1721$

```
      |--------|_____|_____ --------------------------|              o

      ----------------------------------------------------------------------
      0     500   1000   1500    2000    2500    3000    3500    4000
```

center (median) = 1331; bulk of the data between 865 and 1727; range from 452 to 3830; one mild   outlier (3830)

2.35 n = 16 $l_m = (16 + 1)/2 = 8.5$, $l_q = (16 + 2)/4 = 4.5$

median = $(y_{(8)} + y_{(9)})/2 = (4.04 + 4.71)/2 = 4.375$
$Q_1 = (y_{(4)} + y_{(5)})/2 = (3.30 + 3.60)/2 = 3.45$ $Q_3 = (y_{(12)} + y_{(13)})/2 = (5.16 + 5.30)/2 = 5.23$

step = $1.5(Q_3 - Q_1) = 1.5(1.78) = 2.67$
UIF = $5.23 + 2.67 = 7.90$,   UOF = $5.23 + 2(2.67) = 10.57$
LIF = $3.45 - 2.67 = 0.78$,   LOF = $3.45 - 2(2.67) = -1.89$

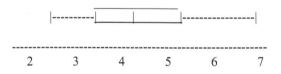

```
      ----------------------------------------------------
      2        3        4        5        6        7
```

center (median) = 4.375; bulk of data between 3.45 and 5.23, data ranges from 22.56 to 6.92; no outliers

2.37

Supplier 1	Supplier 2
n = 12	n = 12

$l_m = (12 + 1)/2 = 6.5$

$l_q = (12 + 2)/4 = 3.5$

median $= (y_{(6)} + y_{(7)})/2 = 85.35$

$Q_1 = (y_{(3)} + y_{(4)})/2 = 84.6$

$Q_3 = (y_{(9)} + y_{(10)})/2 = 86.25$

step $= 1.5(Q_3 - Q_1) = 1.5(1.65) = 2.475$

UIF $= 86.25 + 2.475 = 88.725$

UOF $= 86.25 + 2(2.475) = 91.2$

LIF $= 84.6 - 2.475 = 82.125$

LOF $= 84.6 - 2(2.475) = 79.65$

$l_m = (12 + 1)/2 = 6.5$

$l_q = (12 + 2)/4 = 3.5$

median $= (y_{(6)} + y_{(7)})/2 = 85.25$

$Q_1 = (y_{(3)} + y_{(4)})/2 = 83.65$

$Q_3 = (y_{(9)} + y_{(10)})/2 = 87.6$

step $= 1.5(Q_3 - Q_1) = 1.5(3.95) = 5.925$

UIF $= 87.6 + 5.925 = 93.525$

UOF $= 87.6 + 2(5.925) = 99.45$

LIF $= 83.65 - 5.925 = 77.725$

LOF $= 83.65 - 2(5.925) = 71.8$

```
                              _____
 1    |----------------|      |      |        |----------------|


                        _____
 2       |------|       |           |           |----|

      -----------------------------------------------------------
      82    83    84    85    86    87    88    89
```

Supplier 1 is centered around 85.35; bulk of data between 84.6 and 86.25; data ranges from 82.6 to 88.4; no outliers.

Supplier 2 is centered around 85.25; bulk of data between 83.65 and 87.6; data ranges from 83.1 to 88.1; no outliers.

Supplier 2 is more variable.

2.39

Period I	Period II
n = 20	n = 20

$l_m = (20 + 1)/2 = 10.5$ $l_m = (20 + 1)/2 = 10.5$

$l_q = (20 + 2)/4 = 5.5$ $l_q = (20 + 2)/4 = 5.5$

median $= (y_{(10)} + y_{(11)})/2 = (5 + 6)/2 = 5.5$ median $= (y_{(10)} + y_{(11)})/2 = (2 + 3)/2 = 2.5$

$Q_1 = (y_{(5)} + y_{(6)})/2 = (4 + 5)/2 = 4.5$ $Q_1 = (y_{(5)} + y_{(6)})/2 = (1 + 2)/2 = 1.5$

$Q_3 = (y_{(15)} + y_{(16)})/2 = (8 + 8)/2 = 8$ $Q_3 = (y_{(15)} + y_{(16)})/2 = (4 + 4)/2 = 4$

step $= 1.5(Q_3 - Q_1) = 1.5(3.5) = 5.25$ step $= 1.5(Q_3 - Q_1) = 1.5(2.5) = 3.75$

UIF $= 8 + 5.25 = 13.25$ UIF $= 4 + 3.75 = 7.75$

UOF $= 8 + 2(5.25) = 18.5$ UOF $= 4 + 2(3.75) = 11.5$

LIF $= 4.5 - 5.25 = -.75$ LIF $= 1.5 - 3.75 = -2.25$

LOF $= 4.5 - 2(5.25) = -6$ LOF $= 1.5 - 2(3.75) = -6$

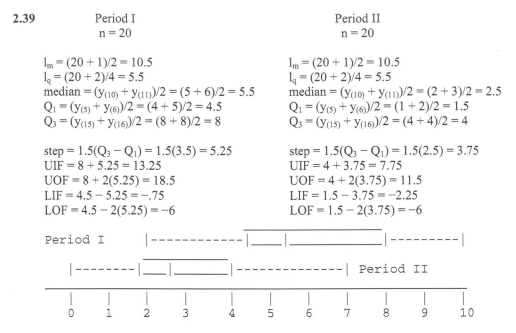

Period I is centered around 5.5; bulk of data between 4.5 and 8; data ranges from 2 to 10; no outliers.
Period II is centered around 2.5; bulk of data between 1.5 and 4; data ranges from 0 to 7; no outliers.
Period II has an overall lower number of industrial accidents than Period I.

2.41

Period I	Period II
n = 15	n = 15

$l_m = (15 + 1)/2 = 8$ $l_m = (15 + 1)/2 = 8$

$l_q = (15 + 3)/4 = 4.5$ $l_q = (15 + 3)/4 = 4.5$

median $= y_{(8)} = 33.5$ median $= y_{(8)} = 34.6$

$Q_1 = (y_{(4)} + y_{(5)})/2 = (33.2 + 33.3)/2 = 33.25$ $Q_1 = (y_{(4)} + y_{(5)})/2 = (33.5 + 34.3)/2 = 33.9$

$Q_3 = (y_{(11)} + y_{(12)})/2 = (33.8 + 33.8)/2 = 33.8$ $Q_3 = (y_{(11)} + y_{(12)})/2 = (34.7 + 34.8)/2 = 34.75$

step $= 1.5(Q_3 - Q_1) = 1.5(.55) = .825$ step $= 1.5(Q_3 - Q_1) = 1.5(.85) = 1.275$

UIF $= 33.8 + .825 = 34.625$ UIF $= 34.75 + 1.275 = 36.025$

UOF $= 33.8 + 2(.825) = 35.45$ UOF $= 34.75 + 2(1.275) = 37.30$

LIF $= 33.25 - .825 = 32.425$ LIF $= 33.9 - 1.275 = 32.625$

LOF $= 33.25 - 2(.825) = 31.60$ LOF $= 33.9 - 2(1.275) = 31.35$

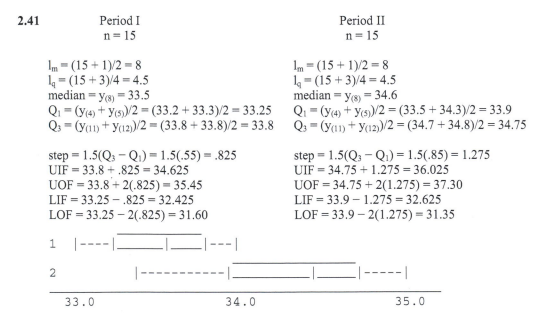

Time Period I is centered around 33.5; bulk of data between 33.25 and 33.8; data ranges from 33.0 to 34.0; no outliers

Time Period 2 is centered around 34.6; bulk of data between 33.9 and 34.75; data ranges from 33.3 to 35.0; no outliers.

In general, Time Period 2 has higher viscosity values and more variability than Time Period 1.

2.43

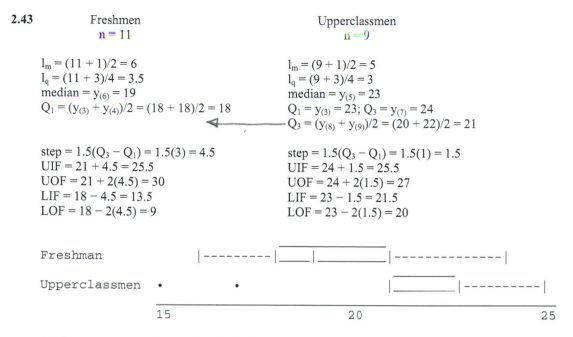

Freshmen
n = 11

$l_m = (11 + 1)/2 = 6$
$l_q = (11 + 3)/4 = 3.5$
median $= y_{(6)} = 19$
$Q_1 = (y_{(3)} + y_{(4)})/2 = (18 + 18)/2 = 18$

step $= 1.5(Q_3 - Q_1) = 1.5(3) = 4.5$
UIF $= 21 + 4.5 = 25.5$
UOF $= 21 + 2(4.5) = 30$
LIF $= 18 - 4.5 = 13.5$
LOF $= 18 - 2(4.5) = 9$

Upperclassmen
n = 9

$l_m = (9 + 1)/2 = 5$
$l_q = (9 + 3)/4 = 3$
median $= y_{(5)} = 23$
$Q_1 = y_{(3)} = 23; Q_3 = y_{(7)} = 24$
$Q_3 = (y_{(8)} + y_{(9)})/2 = (20 + 22)/2 = 21$

step $= 1.5(Q_3 - Q_1) = 1.5(1) = 1.5$
UIF $= 24 + 1.5 = 25.5$
UOF $= 24 + 2(1.5) = 27$
LIF $= 23 - 1.5 = 21.5$
LOF $= 23 - 2(1.5) = 20$

Freshmen are centered around 19; bulk of data between 18 and 21; data ranges from 16 to 24; no outliers.

Upperclassmen are centered around 23; bulk of data between 23 and 24; data ranges from 15 to 25, two extreme outliers (15, 17)

The homework grades for upperclassmen tend to be higher than that of freshmen.
The grades for Freshmen have more variability.

2.45 N = 35   Median = 0.379
Quartiles = 0.378, 0.379

Decimal point is 3 places to the left of the colon

```
    1      1     376 : 0
    7      6     377 : 000000
   16      9     378 : 000000000
          14     379 : 00000000000000
    5      4     380 : 0000
    1      1     381 : 0
```

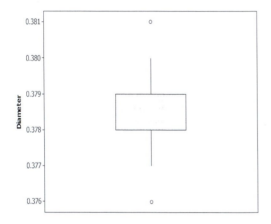
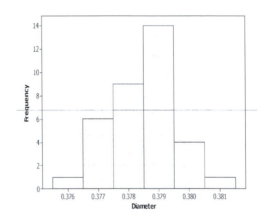

The distribution of the outside diameters is slightly skewed to the left and single peaked with two possible outliers.

We lose the individual data values in the histogram. We prefer stem-and-leaf displays for small data sets since we do not lose the individual data values. For larger data sets, we prefer histograms since we can scale them to fit on our page or screen without losing any further information. The boxplot can identify outliers better than other displays and can show the range where most of the data are clustered together. However, the individual data values are lost in a boxplot.

2.47 N = 35   Median = 0.76
       Quartiles = 0.69, 0.79

Decimal point is 1 place to the left of the colon

```
 1    1   6 : 3
 9    8   6 : 67788899
16    7   7 : 1122244
     11   7 : 56666777899
 8    5   8 : 00113
 3    3   8 : 667
```

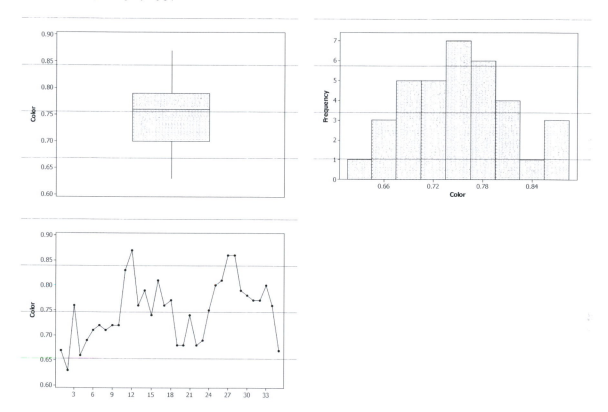

The distribution of the color property is roughly symmetric and single peaked with no possible outliers. The data shows from the timeplot that the property values get higher during the later process. There appears to be some clustering.

We lose the individual data values in the histogram. We prefer stem-and-leaf displays for small data sets since we do not lose the individual data values. For larger data sets, we prefer histograms since we can scale them to fit on our page or screen without losing any further information. The boxplot can identify outliers better than other displays and can show the range where most of the data are clustered together. However, the individual data values are lost in a boxplot. The timeplot can show if there is any changes over time.

2.49 N = 14   Median = 343.3
Quartiles = 343.1, 343.4

Decimal point is at the colon

```
4     3     343  :  011
      5     343  :  33333
5     3     343  :  445
2     1     343  :  7
1     1     343  :  8
```

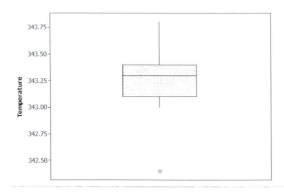 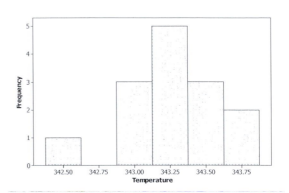

The distribution of the melting points is approximately symmetric and single peaked with one mild outlier.

We lose the individual data values in the histogram. We prefer stem-and-leaf displays for small data sets since we do not lose the individual data values. For larger data sets, we prefer histograms since we can scale them to fit on our page or screen without losing any further information. The boxplot can identify outliers better than other displays and can show the range where most of the data are clustered together. However, the individual data values are lost in a boxplot.

2.51 N = 25   Median = 0.97
Quartiles = 0.94, 0.99

Decimal point is 1 place to the left of the colon

Low: 0.80 0.81

```
      3     1    8 : 9
      3     0    9 :
      5     2    9 : 23
      9     4    9 : 4455
            5    9 : 66677
     11    11    9 : 89999999999
```

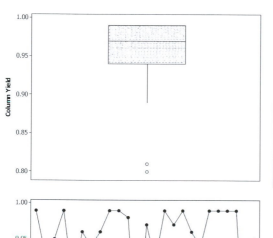

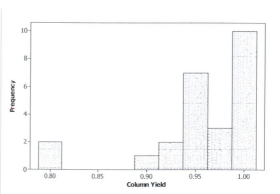

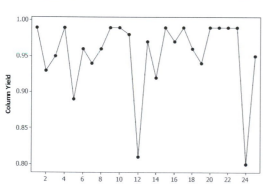

The distribution of the time required to stock a machine is skewed to the left and single peaked with two mild outliers. There is some indication of a cyclical pattern from the time-plot.

We lose the individual data values in the histogram. We prefer stem-and-leaf displays for small data sets since we do not lose the individual data values. For larger data sets, we prefer histograms since we can scale them to fit on our page or screen without losing any further information. The boxplot can identify outliers better than other displays and can show the range where most of the data are clustered together. However, the individual data values are lost in a boxplot.

2.53 Stem-and-leaf of Silica   N   = 22
    Leaf Unit = 1.0

```
    1    2   0
    4    2   222
    4    2
   10    2   677777
   (2)   2   89
   10    3   01
    8    3   2333333
    1    3   4
```

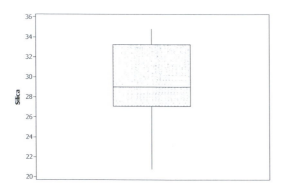
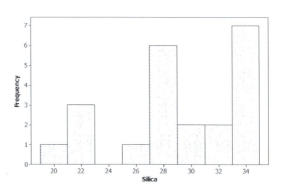

The distribution of the percentage of silica is double peaked with no outliers. We lose the individual data values in the histogram. We prefer stem-and-leaf displays for small data sets since we do not lose the individual data values. For larger data sets, we prefer histograms since we can scale them to fit on our page or screen without losing any further information. The boxplot can identify outliers better than other displays and can show the range where most of the data are clustered together. However, the individual data values are lost in a boxplot.

2.55 Stem-and-leaf of Failure Time N = 84
   Leaf Unit = 1.0

```
21   0   111223333444556667889
29   1   00145679
32   2   889
36   3   0236
39   4   024
(7)  5   0124568
38   6   345
35   7   01589
30   8   4689
26   9   0127
22  10   034456
16  11   024578
10  12   0124568999
```

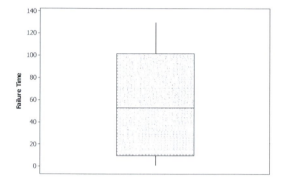

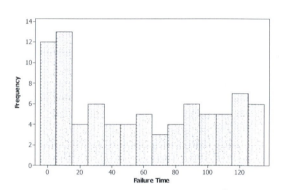

The distribution of failure times has a large peak at the beginning, a possible peak at the end, and is fairly
constant in between. We lose the individual data values in the histogram. We prefer stem-and-leaf displays for
small data sets since we do not lose the individual data values. For larger data sets, we prefer histograms since
we can scale them to fit on our page or screen without losing any further information. The boxplot can identify
outliers better than other displays and can show the range where most of the data are clustered together.
However, the individual data values are lost in a boxplot.

2.57 Stem-and-leaf of Strength   N = 63
Leaf Unit = 0.010

```
  1     5    5
  1     6
  3     7    47
  5     8    14
  6     9    3
  7    10    4
  9    11    13
 14    12    45789
 17    13    069
 22    14    28899
(10)    15    0012345589
 31    16    0111122346667889
 15    17    0036678
  8    18    12449
  3    19
  3    20    01
  1    21
  1    22    4
```

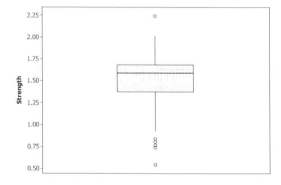

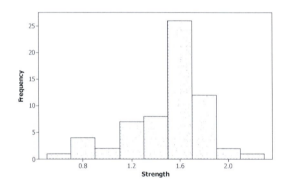

The distribution of glass fiber strengths is skewed left with several outliers. We lose the individual data values in the histogram. We prefer stem-and-leaf displays for small data sets since we do not lose the individual data values. For larger data sets, we prefer histograms since we can scale them to fit on our page or screen without losing any further information. The boxplot can identify outliers better than other displays and can show the range where most of the data are clustered together. However, the individual data values are lost in a boxplot.

2.59 Stem-and-leaf of Stress   N = 16
Leaf Unit - 0.10

```
2   2   59
4   3   23
7   3   677
8   4   0
8   4   77
6   5   113
3   5   6
2   6   0
1   6   9
```

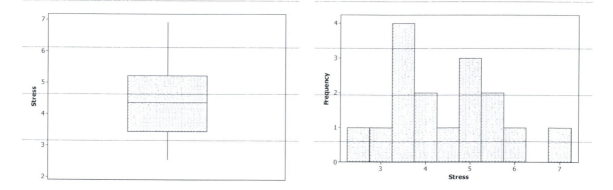

The distribution of bar stresses is skewed right with no outliers. We lose the individual data values in the histogram. We prefer stem-and-leaf displays for small data sets since we do not lose the individual data values. For larger data sets, we prefer histograms since we can scale them to fit on our page or screen without losing any further information. The boxplot can identify outliers better than other displays and can show the range where most of the data are clustered together. However, the individual data values are lost in a boxplot.

2.61

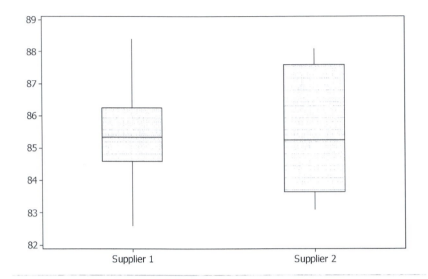

It does not appear that the suppliers are significantly different with regard to center since the boxes overlap. However, this plot does show that Supplier 2 is more variable.

Parallel boxplots compare the differences among two or more groups of data in terms of minimum, maximum, median, range, and variability. It gives us an idea of how they are different.

2.63

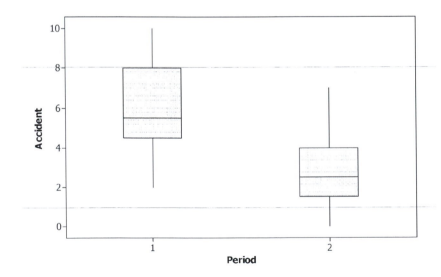

The second five year period has lower rate of accidents as shown in the parallel boxplots.

Parallel boxplots compare the differences among two or more groups of data in terms of minimun, maximum, median, range, and variability. It gives us an idea of how they are different.

2.65

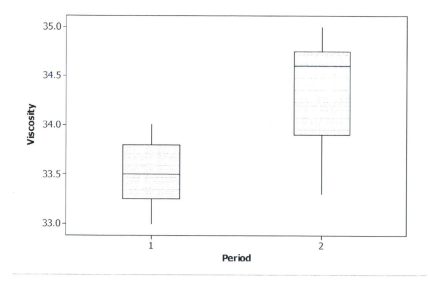

The viscosity of time period 1 has lower values and lower variability than time period 2 as shown in the parallel boxplots.

Parallel boxplots compare the differences among two or more groups of data in terms of minimun, maximum, median, range, and variability. It gives us an idea of how they are different.

2.67

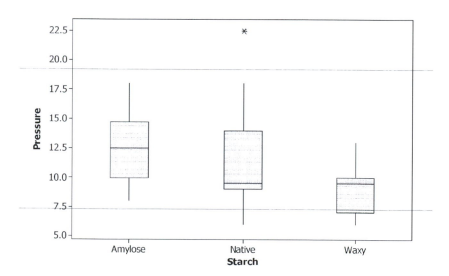

In general, the Waxy Maize starch has lower injection pressure than the other two starches.

Parallel boxplots compare the differences among two or more groups of data in terms of minimun, maximum, median, range, and variability. It gives us an idea of how they are different.

2.69 We eliminate the variability among the blends by looking at the observed differences. It enables us to see the differences between methods easier.

difference = method 1 − method 2
N = 32   Median = −7.75
Quartiles = −9.5, −5.2

Decimal point is 1 place to the right of the colon

```
    1      1    -1 : 2
    8      7    -1 : 1111000
          10    -0 : 9988888888
   14      6    -0 : 776666
    8      3    -0 : 544
    5      4    -0 : 3222
    1      1    -0 : 1
```

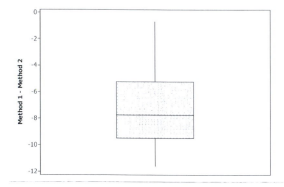
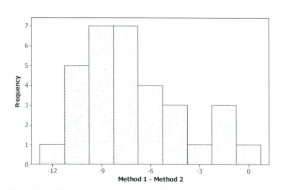

The distribution of the differences between two methods is slightly skewed to the right and single peaked with no outliers. Since all of the differences are below zero, we can conclude that method 1 has significantly lower measurement than method 2.

We lose the individual data values in the histogram. We prefer stem-and-leaf displays for small data sets since we do not lose the individual data values. For larger data sets, we prefer histograms since we can scale them to fit on our page or screen without losing any further information. Boxplot can identify outliers better than other displays and can show the range where most of the data are clustered together. However, the individual data values are lost in a boxplot.

2.71 We eliminate the variability among the plates by looking at the observed differences. This enables us to see the differences between location easier.

difference = Location A − Location B
N = 10   Median = 0.425
Quartiles = 0, 1.35

Decimal point is at the colon

```
    1   1   -0 : 7
    1   0   -0 :
    5   4    0 : 0011
    5   2    0 : 77
    3   2    1 : 33
    1   0    1 :
    1   1    2 : 0
```

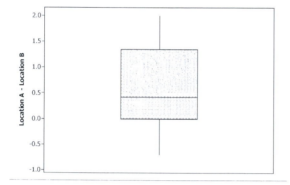
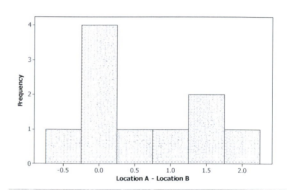

The distribution of the differences between the two locations is roughly symmetric with no outliers. Since the differences are all positive except one, it strongly suggests that location A on the plates is consistently thicker than location B as the mechanical engineer believes.

We lose the individual data values in the histogram. We prefer stem-and-leaf displays for small data sets since we do not lose the individual data values. For larger data sets, we prefer histograms since we can scale them to fit on our page or screen without losing any further information. Boxplot can identify outliers better than other displays and can show the range where most of the data are clustered together. However, the individual data values are lost in a boxplot.

2.73

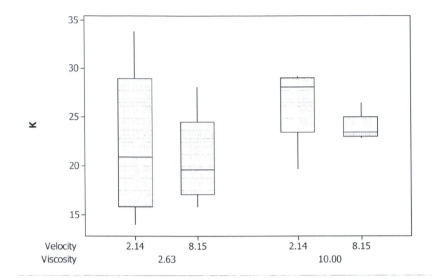

There is more variation at a viscosity of 2.63 than at a viscosity of 10. Also, there is more variation at a velocity of 2.14 than at a velocity of 8.15. There is some evidence that K is higher using a velocity of 10.

CHAPTER 3

MODELING RANDOM BEHAVIOR

3.1 **a** $\Pr(Y < 2) = \Pr(Y = 0) + \Pr(Y = 1) = .512 + .384 = .896$

b $\Pr(Y > 1) = \Pr(Y = 2) + \Pr(Y = 3) = .096 + .008 = .104$

c $\mu = \Sigma_i \, y_i \, p(y_i)$
$\mu = 0(.512) + 1(.384) + 2(.096) + 3(.008)$
$\mu = 0 + .384 + .192 + .024$
$\mu = .60$

d $\sigma^2 = \Sigma_i \, y_i^2 \, p(y_i) - \mu^2$
$\sigma^2 = 0(.512) + 1(.384) + 4(.096) + 9(.008) - (.6)^2$
$\sigma^2 = .84 - .36$
$\sigma^2 = .48$
$\sigma = .693$

3.3 **a** $\Pr(Y \le 2) = \Pr(Y = 0) + \Pr(Y = 1) + \Pr(Y = 2) = .005 + .010 + .035 = .05$

b $\Pr(Y \ge 1) = 1 - \Pr(Y < 1) = 1 - \Pr(Y=0) = 1 - .005 = .995$

c $\mu = \Sigma_i \, y_i \, p(y_i) = 0(.005) + 1(.010) + 2(.035) + 3(.050) + 4(.900) = 3.83$

d $\sigma^2 = \Sigma_i \, y_i^2 \, p(y_i) - \mu^2$
$\sigma^2 = 0(.005) + 1(.010) + 4(.035) + 9(0.50) + 16(.900) - (3.83)^2$
$\sigma^2 = 15 - (3.83)^2$
$\sigma^2 = .3311$
$\sigma = .5754$

3.5 **a** $\Pr(Y = 2) = .2098$

b $\Pr(Y \ge 1) = 1 - \Pr(Y < 1) = 1 - \Pr(Y = 0) = 1 - .3106 = .6894$

c $E(Y) = \Sigma_i\, y_i\, p(y_i)$
$E(Y) = 0(.3106) + 1(.4313) + 2(.2098) + 3(.0442) + 4(.0040) + 5(.0001)$
$E(Y) = 0 + .4313 + .4196 + .1326 + .0160 + .0005$
$E(Y) = 1.0000 = \mu$

$\sigma^2 = \Sigma_i\, y_i^2\, p(y_i) - \mu^2$
$\sigma^2 = 0(.3106) + 1(.4313) + 4(.2098) + 9(.0442) + 16(.0040) + 25(.0001) - (1)^2$
$\sigma^2 = 1.7348 - 1$
$\sigma^2 = .7348$
$\sigma = .8572$

3.7 **a** $\Pr(Y \le 1) = \Pr(Y = 0) + \Pr(Y = 1) = .716 + .180 = .896$

b $\Pr(1 < Y < 5) = P(Y = 2) + P(Y = 3) + P(Y = 4) = .060 + .020 + .010 = .090$

c $\mu = E(Y) = \Sigma_i\, y_i\, p(y_i)$
$\mu = 0(.716) + 1(.180) + 2(.060) + 3(.020) + 4(.010) + 5(.010) + 6(.002) + 7(.000) + 8(.002)$
$\mu = .478$

d $\sigma^2 = E(Y^2) - \mu^2 = \Sigma_i\, y_i^2\, p(y_i) - [E(Y)]^2$
$\sigma^2 = 0(.716) + 1(.180) + 4(.060) + 9(.020) + 16(.010) + 25(.010) + 36(.002) + 49(.000) + 64(.002) - (.478)^2$
$\sigma^2 = 1.21 - .228$
$\sigma^2 = .9815$
$\sigma = .9907$

3.9 p = probability a drawer gets stuck = .02
n = 10
Y = number of drawers that get stuck
Y is distributed as Binomial ($n = 10$, $p = .02$)

a $\Pr(Y = 0)$ $= \dbinom{10}{0}(.02)^0(.98)^{10}$

$= .8171$

b $\Pr(Y \ge 1)$ $= 1 - P(Y = 0)$
$= 1 - .8171$
$= .1829$

c $E(Y) = np = 10(.02) = .2$
$\sigma^2 = npq = 10(.02)(.98) = .196$
$\sigma = .4427$

3.11 p = probability a filter leaks = .05
  n = 5
  Y = number of filters that leak
  Y is distributed as Binomial (n = 5, p = .05)

a $\Pr(Y = 5)$ $=$ $\binom{5}{5}(.05)^5(.95)^0$

 $=$ $.0000003$

b $\Pr(1 \le Y \le 3)$ $=$ $P(Y = 1) + P(Y = 2) + P(Y = 3)$

 $=$ $\binom{5}{1}(.05)^1(.95)^4 + \binom{5}{2}(.05)^2(.95)^3 + \binom{5}{3}(.05)^3(.95)^2$

 $=$ $.2036 + .0214 + .0011 = .2261$

c $E(Y) = np = 5(.05) = .25$
  $\sigma^2 = npq = 5(.05)(.95) = .2375$
  $\sigma = .4873$

3.13 a n = 4 pumps used by low-pressure coolant injection systems.
  p = probability of a pump failing to run = .16
  Y = number of pumps failing to run out of 4.
  Y is distributed as binomial (n = 4, p = .16)

 i $\Pr(Y = 4)$ $=$ $\binom{4}{4}(.16)^4(1 - .16)^{4-4} = (1)(.000655)(1) = .000655$

 ii $\Pr(Y \ge 1)$ $=$ $1 - \Pr(Y < 1)$
   $=$ $1 - \Pr(Y = 0)$
   $=$ $1 - \binom{4}{0}(.16)^0(.84)^4$
   $=$ $1 - .4979$
   $=$ $.5021$

 iii $E(Y) = np = 4(.16) = .64$
   $\sigma^2 = npq = 4(.16)(.84) = .5376$
   $\sigma = .7332$

b n = 8 pumps used by low-pressure coolant injection systems.
  p = .16
  Y = number of pumps failing to run out of 8.
  Y is distributed as binomial (n = 8, p = .16)

 i $\Pr(Y = 0)$ $=$ $\binom{8}{0}(.16)^0(.84)^8 = .2479$

 ii $\Pr(Y = 2)$ $=$ $\binom{8}{2}(.16)^2(.84)^6 = .2518$

 iii $E(Y) = np = 8(.16) = 1.28$
   $\sigma^2 = npq = 8(.16)(.84) = 1.0752$
   $\sigma = 1.037$

3.15 n = 15 cars
p = probability that a car requires service under the warranty = .07.
Y = number of cars that require service under warranty out of 15 cars.
Y   is distributed as binomial (n = 15, p = .07)

a $Pr(Y = 1)$ = $\binom{15}{1}(.07)^1(.93)^{15-1} = .3801$

b $Pr(Y > 1)$ = $1 - Pr(Y \le 1)$
= $1 - Pr(Y = 0) - Pr(Y = 1)$
= $1 - \binom{15}{0}(.07)^0(.93)^{15} - Pr(Y = 1)$

= $1 - .3367 - .3801$
= $.2832$

c $E(Y) = np = 15(.07) = 1.05$
$\sigma^2 = npq = 15(.07)(.93) = .9765$
$\sigma = .9882$

3.17 n = 8 batches dyed by the process
p = probability of a batch being *rejected* = .05
Y = number of batches being *rejected* out of 8.
Y   is distributed as binomial (n = 8, p = .05)

a $Pr(Y \le 1)$ = $Pr(Y = 0) + Pr(Y = 1)$
= $\binom{8}{0}(.05)^0(.95)^8 + \binom{8}{1}(.05)^1(.95)^7$

= $.6634 + .2793$
= $.9428$

b $Pr(Y \ge 1)$ = $1 - Pr(Y < 1)$
= $1 - Pr(Y = 0)$
= $1 - \binom{8}{0}(.05)^0(.95)^8$

= $1 - .6634$
= $.3366$

c X = number of batches being accepted
X is distributed as binomial (n = 8, p = .95)

$E(X) = 8(.95) = np = 7.6$
$\sigma^2 = 8(.95)(.05) = npq = .38$
$\sigma = .6164$

d $E(Y)$ $= np = 8(.05) = .4$
$\sigma^2 = 8(.05)(.95) = .38$
$\sigma = .6164$

e The expected values differ but the variances and standard deviations are the same.
Actually there is only one response since Y = 8 − X and Pr(rejected) = 1 − Pr(accepted).

3.19 n = 15 batches of an aluminum alloy
  p = probability that a batch does not meet the new specification = .2
  Y = number of batches that fail to meet the new specification out of 15.
  Y   is distributed as binomial (n = 15, p = .2)

a Pr(Y = 3)   = $\binom{15}{3}(.2)^3(.8)^{12} = (455)(.008)(.0687) = .2501$

b Pr(Y = 0)   = $\binom{15}{0}(.2)^0(.8)^{15} = .0352 = $ Pr(all meet specs)

c E(Y) = np = 15(.2) = 3
  σ^2 = npq = 15 (.2)(.8) = 2.4
  σ = 1.5492

d X = number of batches that meet the new specification out of 15.
  X is distributed as binomial (n = 15, p = .8)

E(X) = np = 15(.8) = 12
σ^2 = npq = 15(.8)(.2) = 2.4
σ = 1.5492

e The expected values differ but the variances and standard deviations are the same.
  There is only one response variable since X = 15 − Y and Pr(fail) = 1 − Pr(pass).

3.21 Y = number of accidents for any given month.
  Y is distributed as Poisson (λ = 1.5)

a Pr(Y = 0) = $1.5^0 e^{-1.5}/0! = .2231$

b Pr(Y ≥ 1) = 1 − Pr(Y < 1) = 1 − Pr(Y = 0) = 1 − .2231 = .7769

c Pr(Y = 5) = $1.5^5 e^{-1.5}/5! = .0141$

d μ = E(Y) = λ = 1.5
  $\sigma^2 = \lambda$ = 1.5
  σ = 1.2247

3.23 Y = number of warranty claims within one year of purchase for a particular system on a single car model.
  Y is distributed as Poisson (λ = .75)

a Pr(Y = 0) = $.75^0 e^{-.75}/0! = .4724$

b Pr(Y = 3) = $.75^3 e^{-.75}/3! = .0332$

c E(Y) = λ = .75
  $\sigma^2 = \lambda$ = .75
  σ = .8660

3.25 Y = number of defects per car
Y is distributed as Poisson ($\lambda = 5$)

  a $\Pr(Y = 7) = 5^7 e^{-5}/7! = 78125(.0067)/5040 = .1044$

  b $\mu = E(Y) = \lambda = 5$
    $\sigma^2 = \lambda = 5$
    $\sigma = 5^{1/2} = 2.2361$

3.27 Y = number of accidents at a busy intersection per month
Y is distributed as Poisson ($\lambda = 2.5$)

  a $\Pr(Y = 0)$ $=$ $2.5^0 e^{-2.5}/0!$ $=$ $.0821$

  b $\Pr(Y > 1)$ $=$ $1 - \Pr(Y \le 1)$
                     $=$ $1 - \Pr(Y = 0) - \Pr(Y = 1)$
                     $=$ $1 - (2.5^0 e^{-2.5}/0!) - (2.5^1 e^{-2.5}/1!)$
                     $=$ $1 - .0821 - .2052$
                     $=$ $.7127$

  c $\mu = E(Y) = \lambda = 2.5$
    $\sigma^2 = \lambda = 2.5$
    $\sigma = 1.5811$

3.29 Y = number of defects per 100 pages of text
Y is distributed as Poisson ($\lambda = 1$) or ($\lambda = .6$)
$$P(y) = \frac{(\lambda)^y \exp(-\lambda)}{y!} \quad y = 0,1,2,\ldots$$

  a $\Pr(Y = 0) = 1^0 e^{-1}/0! = .3679$

  b $\Pr(Y = 0) = .6^0 e^{-.6}/0! = .5488$

  c the current machine does not satisfy their criteria but the new machine does

  d $\mu = \lambda = 1$
    $\sigma^2 = \lambda = 1$
    $\sigma = 1$

3.31 Y = number of applicants interviewed until one has training
Pr(success) = .30 = p
Pr(failure) = .70 = 1 − p
Y follows a geometric distribution, so $p(y) = (1 - .3)^{y-1} .3$ for y = 1,2,....

  a $\Pr(Y = 4) = (1 - .3)^{4-1} .3 = .1029$

b $\Pr(Y = 1) = (1 - .3)^{1-1}.3 = .3$

c $E(Y) = 1/p = 1/.3 = 3.333$
$\sigma^2 = (1 - p)/p^2 = (1 - .3)/.3^2 = 7.7778$
$\sigma = 2.789$

d X = number of applicants until second has training
Assume X follows a negative binomial distribution with r = 2 and p = .3

$$\Pr(X = 7) = \binom{7-1}{2-1}(.3)^2(.7)^{7-2}$$
$$= \binom{6}{1}(.3)^2(.7)^5$$
$$= .0908$$

3.33 Y = number of parts made until the first part scrapped.
Pr(parts scrapped) = .10 = p
Y is distributed as geometric with p = .10, so $\Pr(Y = y) = (1 - p)^{y-1}p = (.9)^{y-1}(.1)$ for y = 1,2, $\cdots$

a $\Pr(Y = 10) = (.9)^9(.1) = .0387$

b $\Pr(Y \geq 2) = 1 - \Pr(Y < 2) = 1 - \Pr(Y = 1) = 1 - (.9)^0(.1) = 1 - .1 = .90$

c $E(Y) = 1/p = 1/.10 = 10$
$\sigma^2 = (1 - p)/p^2 = (1 - .1)/(.1)^2 = .9/.01 = 90$
$\sigma = 9.4868$

d X = number of parts made until the fourth part scrapped
X is distributed as negative binomial (p = .10, r = 4)

$$\Pr(X = 20) = \binom{20-1}{4-1}(.1)^4(.9)^{20-4}$$
$$= \binom{19}{3}(.1)^4(.9)^{16}$$
$$= 969(.1)^4(.9)^{16}$$
$$= .0180$$

e $E(X) = r/p = 4/.10 = 40$
$\sigma^2 = r(1 - p)/p^2 = 4(.9)/(.1)^2 = 360$
$\sigma = 18.9737$

3.35 Y = number of automobiles sold until the first claim.
Pr(autos have required service under warrant) = .07
Y is distributed as geometric (p = .07)
$\Pr(Y = y) = (1 - p)^{y-1}p = (.93)^{y-1}(.07)$ for y = 1,2, $\cdots$

a $\Pr(Y = 10) = (.93)^9(.07) = .0364$

b E(Y) = 1/p = 1/.07 = 14.2857
σ^2 = .93/.07^2 = 189.8
σ = 13.777

c X = number of autos sold until the third car requires service.
X is distributed as negative binomial with p = .07, r = 3

$$\text{Pr}(X = x) \quad = \quad \binom{x-1}{r-1}(p)^r(1-p)^{x-r} \quad \text{for } x = r, r+1, \cdots$$

$$\text{Pr}(X = x) \quad = \quad \binom{x-1}{2}(.07)^3(.93)^{x-3} \quad \text{for } x = 3, 4, \cdots$$

$$\text{Pr}(X = 20) \quad = \quad \binom{19}{2}(.07)^3(.93)^{17}$$

$$= \quad 171\,(.07)^3(.93)^{17}$$
$$= \quad .0171$$

d E(X) = r/p = 3/.07 = 42.8571
σ^2 = r(1 − p)/p^2 = 3(.93)/(.07)^2 = 569.3878
σ = 23.8618

3.37 P(select a winning slip of paper) = .02
Y is distributed as hypergeometric (N =100, r = 2, n = 4)

a $\text{Pr}(Y = 1) = \dfrac{\binom{2}{1}\binom{98}{3}}{\binom{100}{4}} = .0776$

b $\text{Pr}(Y = 0) = \dfrac{\binom{2}{0}\binom{98}{4}}{\binom{100}{4}} = .9212$

c $\text{E}(Y) = \dfrac{4(2)}{100} = .08$

$\sigma^2 = \dfrac{4(2)(100-2)(100-4)}{(100)^2(100-1)} = .076$

$\sigma = \sqrt{.076} = .2757$

3.39 Y is distributed as exponential (λ = .05 failures per million revolutions)
f(y) = .05 e$^{-.05y}$ if y > 0 and zero otherwise

a $\int_0^{.5} .05\,e^{-.05y}\,dy = 1 - e^{-.05(.5)} = .0247$

b $\int_3^{\infty} .05\,e^{-.05y}\,dy = e^{-.05(3)} - 0 = .8607$

c E(Y) = 1/λ = 1/.05 = 20

d Var(Y) = σ^2 = 1/λ^2 = 1/(.05)^2 = 400.
σ = 20

3.41 Y is distributed as exponential ($\lambda = .03$)
$f(y) = .03\, e^{-.03y}$ if $y > 0$ and zero otherwise

a $\Pr(Y < 20) = \int_0^{20} .03\exp(-.03y)\,dy = 1 - e^{-.03(20)} = .4512$

b $\Pr(Y > 30) = \int_{30}^{\infty} .03\exp(-.03y)\,dy = e^{-.03(30)} - 0 = .4066$

c $E(Y) = 1/\lambda = 1/.03 = 33.333$

d $\mathrm{Var}(Y) = \sigma^2 = 1/\lambda^2 = 1/(.03)^2 = 1111.111$
$\sigma = 33.333$

3.43 a $F(y) = \int_a^y \dfrac{1}{b-a}\,dw = \dfrac{w}{b-a}\Big|_a^y$

$F(y) = 0 \qquad\qquad\quad$ if $y < a$
$F(y) = (y - a)/(b - a) \qquad$ if $a \le y \le b$
$F(y) = 1 \qquad\qquad\quad$ if $y > b$

b $E(Y) = \int_a^b \dfrac{w}{b-a}\,dw = \dfrac{w^2}{2(b-a)}\Big|_a^b = \dfrac{b^2 - a^2}{2(b-a)} = \dfrac{(b-a)(b+a)}{2(b-a)} = \dfrac{a+b}{2}$

c $E(Y^2) = \int_a^b \dfrac{w^2}{b-a}\,dw = \dfrac{w^3}{3(b-a)}\Big|_a^b = \dfrac{b^3 - a^3}{3(b-a)} = \dfrac{(b^2 + ab + a^2)}{3}$

$\sigma^2 = \mathrm{Var}(Y) = E(Y^2) - (E(Y))^2 \qquad = \dfrac{(b^2 + ab + a^2)}{3} - \left(\dfrac{b+a}{2}\right)^2$

$\qquad = \dfrac{4(b^2 + ab + a^2)}{12} - \dfrac{3(b^2 + 2ab + a^2)}{12}$

$\qquad = \dfrac{b^2 + 2ab + a^2}{12}$

$\qquad = (b-a)^2/12$

$\sigma \qquad = (b-a)/(12^{1/2})$

d Y = time accident occurs during month
Y is distributed as uniform (a = 0, b = 30)

(i) $\Pr(0 \le Y \le 15) = F(15) = (15 - 0)/(30 - 0) = 1/2$

(ii) $\Pr(Y \ge 25) = 1 - P(Y \le 25) = 1 - F(25) = 1 - (25 - 0)/(30 - 0) = 5/30 = 1/6$

(iii) $E(Y) = (30 + 0)/2 = 15$ days

(iv) $\sigma^2 = (30 - 0)^2/12 = 75$ days
$\sigma = 8.66$ days

3.45 Y is distributed as Weibull ($\lambda = .4, \beta = 2$)

a $F(Y) = \int_0^y \lambda\beta(\lambda y)^{\beta - 1} \exp[-(\lambda y)^\beta]dx = -\exp[-(\lambda y)^\beta]\big|_0^y = 1 - \exp[-(\lambda y)^\wedge\beta] = 1 - \exp[-(.4y)^2]$
$F(Y)$ simplifies to $1 - \exp[-.16y^2]$

b $\Pr(Y \le 1.2) = F(1.2) = 1 - \exp[-.16(1.2)^2] = 1 - \exp(-.2304) = .2058$

c $E(Y) = \Gamma(1 + 1/\beta)\lambda^{-1} = \Gamma(1 + 1/2)(.4)^{-1} = (\pi^{1/2}/2)\,2.5 = (5/4) * \pi^{1/2}$

d $\sigma^2 = [\Gamma(1 + 2/\beta) - (\Gamma(1 + 1/\beta))^2]\lambda^{-2} = [\Gamma(2) - (\pi/4)](.4)^{-2} = [1 - (\pi/4)]6.25 = 1.34$
$\sigma = 1.16$

e 0* 4
0●8 9
1* 0 1 1 2 2 2 3 4 4
1●6 6 6 6 6 6 7 7 7 8 8 8 9
2* 0 0 0 1 1 2 2
2●5 5 8 8 8 9
3* 0 2
3●5 7 7 7 7
4* 4
4●7 9
5* 1
5●6

Skewed right, single peaked around 1.5-2.0

3.47 Y = thicknesses of bolts
Y is distributed as Normal ($\mu = 10.0$mm, $\sigma = 1.6$mm)

a $\Pr(9.2 \le Y \le 10.8)$

$\qquad\qquad = \quad \Pr[(9.2 - 10)/1.6 \le (Y - \mu)/\sigma \le (10.8 - 10)/1.6]$
$\qquad\qquad = \quad \Pr(-.5 \le Z \le .5)$
$\qquad\qquad = \quad \Pr(Z < .5) - \Pr(Z < -.5)$
$\qquad\qquad = \quad .1915 + .1915$
$\qquad\qquad = \quad .3830$

b $\Pr(Y \le 9.2) = \Pr[(Y - \mu)/\sigma \le (9.2 - 10)/1.6] = \Pr(Z \le -.5) = .3085$

3.49 Y = width of part

Y is distributed as Normal ($\mu = 100$, $\sigma = 8$)

a $\Pr(Y > 110) = \Pr[(Y - \mu) / \sigma > (110 - 100)/8] = \Pr(Z > 1.25) = .1056$

b $\Pr(Y < 90) = \Pr[(Y - \mu)/\sigma < (90 - 100)/8] = \Pr(Z < -1.25) = .1056$

c $\Pr(99 < Y < 101) = \Pr[(99 - 100)/8 < (Y - \mu) / \sigma < (101 - 100)/8]$
$$= \Pr(-.125 < Z < .125)$$
$$= .5517 - .4483$$
$$= .1034$$

d $\Pr(Y > 101) < .10$
$\Pr[(Y - \mu)/\sigma < (101 - \mu)/8] > .10$ (switching direction of both inequalities)
$(101 - \mu)/8 = 1.28$ because we know $\Pr(Z > 1.28) = .10$
$\mu = 90.76$

3.51 Y = percentage of butterfat in 2 year old cows

Y   is distributed as Normal ($\mu = 4.51$, $\sigma = 3.48$)

a $P(4.25 < Y < 4.6) = P\left(\dfrac{4.25 - 4.51}{.348} < \dfrac{Y - \mu}{\sigma} < \dfrac{4.60 - 4.51}{.348}\right)$
$$= P(-.75 < Z < .26)$$
$$= P(Z < .26) - P(Z < -.75)$$
$$= .6020 - .2275$$
$$= .3745$$

b $P(Y > 4.75) \quad = P\left(\dfrac{Y - \mu}{\sigma} > \dfrac{4.75 - 4.51}{.348}\right)$
$$= P(Z > .69)$$
$$= .2451$$

3.53 Y = volume per cent

Y   is distributed as normal ($\mu = 70$, $\sigma = 2$)

a $\Pr(Y > 75) = \Pr(Z > (75 - 70)/2) = \Pr(Z > 2.5) = 1 - \Pr(Z \le 2.5) = 1 - .9938 = .0062$

b $\Pr(67 < Y < 73) \quad = \quad \Pr[(67 - 70)/2 < (Y - \mu)/\sigma < (73 - 70)/2]$
$$= \quad \Pr(-1.5 < Z < 1.5)$$
$$= \quad \Pr(Z \le 1.5) - \Pr(Z \le 1.5)$$
$$= \quad .9332 - .0668$$
$$= \quad .8664$$

c $\Pr(Y < 70) < .02$
$\Pr[(Y - \mu)/\sigma < (70 - \mu)/2] < .02$
$(70 - \mu)/2 = -2.06$ since $\Pr(Z < -2.06) < .02$
$\mu = 70 - 2(-2.06) = 74.12$

3.55   Y = tons of daily production of a sulfuric acid process.
Y   is distributed as normal ($\mu = 400$, $\sigma^2 = 225$)
Since $\sigma^2 = 225$, $\sigma = 15$

a Pr($375 \leq Y \leq 425$)

	=	Pr[$(375 - 400)/15 \leq (Y - \mu)/\sigma \leq (425 - 400)/15$]
	=	Pr($-1.67 \leq Z \leq 1.67$)
	=	Pr($Z \leq 1.67$) $-$ Pr($Z \leq -1.67$)
	=	.9525 $-$ .0475
	=	.905

b Pr($Y < 360$) = Pr[$(Y - \mu)/\sigma < (360 - 400)/15$] = Pr($Z < -2.67$) = .0038

3.57   Y = width for a part from the plastic injection molding process.
n = 4
Assuming the Central Limit Theorem, $\overline{Y}$ is distributed as Normal ($\mu = 100$, $\sigma = 8/\sqrt{4} = 4$)

a Pr($\overline{Y} < 99$) + Pr($\overline{Y} > 101$) $= \mathrm{Pr}\left(\dfrac{\overline{Y} - \mu}{\sigma/\sqrt{n}} < \dfrac{99 - 100}{8/\sqrt{4}}\right) + \mathrm{Pr}\left(\dfrac{\overline{Y} - \mu}{\sigma/\sqrt{n}} > \dfrac{101 - 100}{8/\sqrt{4}}\right)$
$= \mathrm{Pr}(Z < -1/4) + \mathrm{Pr}(Z > 1/4)$
$= .4013 + (1 - .5987)$
$= .8026$

b Assume that a sample of 4 parts is sufficiently large enough to use the Central Limit Theorem.

3.59   Y = breaking strength
$\overline{Y}$ is distributed as Normal ($\mu = 165, \sigma = 5/\sqrt{12}$)

a Pr($\overline{Y} < 162$)

	=	$P\left(\dfrac{\overline{Y} - \mu}{\sigma/\sqrt{n}} < \dfrac{162 - 165}{5/\sqrt{12}}\right)$
	=	Pr($Z < -2.08$)
	=	.0188

b Pr($164 < \overline{Y} < 166$)

	=	$P\left(\dfrac{164 - 165}{1.4434} < \dfrac{Y - \mu}{\sigma/\sqrt{n}} < \dfrac{166 - 165}{1.4434}\right)$
	=	Pr($-.69 < Z < .69$)
	=	Pr($Z < .69$) $-$ Pr($Z < -.69$)
	=	.7549 $-$ .2451
	=	.5098

c The sample of twelve batches of pellets is large enough to assume the Central Limit Theorem. To check this assumption, we can construct a stem and leaf display to see if it is bell-shaped.

3.61 Y has a Normal distribution with $\mu = .65$, $\sigma = .015$
Assuming the Central Limit Theorem, $\overline{Y}$ is distributed as Normal ($\mu = .65$, $\sigma = .015/\sqrt{10}$)

a $\Pr(\overline{Y} > .67) = \Pr\left(\dfrac{\overline{Y} - \mu}{\sigma/\sqrt{n}} > \dfrac{.67 - .65}{.015/\sqrt{10}}\right) = \Pr(Z > 4.22) = 0$

b $\Pr(\overline{Y} < .64) = \Pr\left(\dfrac{\overline{Y} - \mu}{\sigma/\sqrt{n}} < \dfrac{.64 - .65}{.015/\sqrt{10}}\right) = \Pr(Z < -2.11) = .0174$

c $\Pr(.62 < \overline{Y} < .66) = \Pr\left(\dfrac{.62 - .65}{.015/\sqrt{10}} < \dfrac{\overline{Y} - \mu}{\sigma/\sqrt{n}} < \dfrac{.66 - .65}{.015/\sqrt{10}}\right)$
$= \Pr(-6.32 < Z < 2.11)$
$= .9826$

d The sample of ten batches of polyester polymer is large enough to assume $\overline{Y}$ is approximately Normally distributed by the Central Limit Theorem. To check these assumptions we can construct stem and leaf diagram of ten batches and check to see if mound-shaped with rapidly dying tails.

3.63 a $\overline{y} = (34 + 31 + 37 + 39 + 36)/5 = 177/5 = 35.4$

b Assuming the Central Limit Theorem, $\overline{Y}$ is distributed as Normal ($\mu = 34.2$, $\sigma = 2.8/\sqrt{5}$)

$$
\begin{aligned}
\Pr(\overline{Y} \geq 35.4) \quad &= \quad \Pr\left(\dfrac{\overline{Y} - \mu}{\sigma/\sqrt{n}} \geq \dfrac{35.4 - 34.2}{2.8/\sqrt{5}}\right) \\
&= \quad \Pr(Z \geq .96) \\
&= \quad 1 - .8315 \\
&= \quad .1685
\end{aligned}
$$

c The number furthest from the current sample mean has most influence. Remove 31.
The number closest to the current sample mean has least influence. Remove 36.

d For the distribution of these thicknesses, we assume that the sample mean of 5 follows a normal distribution by the Central Limit Theorem.

e normal probability plot appears to follow a straight line, so we are comfortable with assumptions.

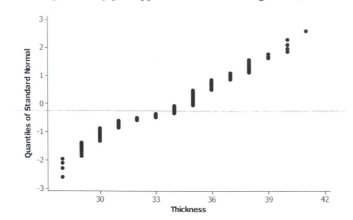

3.65 a $\bar{y} = (.90 + .93 + .95 + .86 + .90 + .87 + .93 + .92)/8 = 7.26/8 = .9075$

b $s^2 = [(.90)^2 + (.93)^2 + (.95)^2 + \cdots + (.92)^2 - (7.26)^2/8]/7 = .00675/7 = .000964$
$s = .03105$

$$t = \frac{\bar{y} - \mu}{\sigma / \sqrt{n}} = \frac{.9075 - .93}{.03105 / \sqrt{8}} = -2.05$$

$|t| = |-2.05| > 2$
It is unlikely, though possible, to observe $t = -2.05$ due to random chance.

c We assumed the distribution of yields is single peaked, relatively symmetric and the tails die rapidly and the sample size of 8 is large enough so that $\dfrac{\bar{y} - \mu}{s / \sqrt{n}}$ follows a t distribution with 7 degrees of freedom.

Stem	Leaves	No.	Depth
.8●	67	2	2
.9*	00233	5	
.9●	5	1	1

We should feel comfortable that the assumptions hold given the stem and leaf diagram.

3.67 a $\bar{y} = (.98 + 101 + 102 + 97 + 101 + 98 + 100 + 92 + 98 + 100)/10 = 987/10 = 98.7$
$s^2 = [(98^2 + 101^2 + \cdots + 100^2) - (987)^2/10]/9 = 8.23$
$s = 2.87$

b $t = \dfrac{\bar{y} - \mu}{\sigma / \sqrt{n}} = \dfrac{98.7 - 100}{2.87 / \sqrt{10}} = -1.43$
$|t| = |-1.43| = 1.43 < 2$. Does not appear to be a rare event.

c   We assumed that the distribution of nominal power is approximately normal. However, the sample of 10 powers does not reflect this assumption. The stem and leaf diagram suggests a population that is skewed left. Hence the assumption of normality is violated.

Stem	Leaves	No.	Depth
9*	2	1	1
9●	7888	4	5
10*	00112	5	5

3.69 **a** $\bar{y} = \Sigma y_i / n = 282/15 = 18.8$

b $t = \dfrac{18.8 - 20}{10.7185 / \sqrt{15}} = -.43$

It is highly possible to observe $t = -.43$ due to random chance.

c We assumed that the distribution of days is single peaked, relatively symmetric and the tails die rapidly.

We also assumed that the sample size of 15 is large enough so that $\dfrac{\bar{y} - \mu}{s / \sqrt{n}}$ follows a t distribution with 14

degrees of freedom.

Stem	Leaves	No.	Depth
1*	011223	6	6
1●	7788	4	
2*	0233	4	5
2●		0	1
3*		0	1
3●		0	1
4*		0	1
4●		0	1
5*	4	1	1

The stem-and-leaf display indicates that the data is single peaked and skewed to the right. These data does not meet the assumptions. There is also a possible outlier at 54.

3.71 **a** $\bar{y} = \Sigma y_i / n = 3709/26 = 142.6538$

$s^2 = \dfrac{n\Sigma y_i^2 - (\Sigma y_i)^2}{n(n-1)} = \dfrac{26(770205) - (3709)^2}{26(25)} = 9644.0754$

$s = \sqrt{s^2} = \sqrt{9644.0754} = 98.2043$

b $t = \dfrac{142.6538 - 220}{98.2043 / \sqrt{26}} = -4.02$; it is rare to observe $t = -4.02$ due to random chance.

c   We assumed that the aluminum contamination has a single peaked, relatively symmetric distribution in which the tails die rapidly. We have also assumed that the sample size of 26 is large enough so that $\dfrac{\overline{y} - \mu}{s / \sqrt{n}}$ follows a t distribution with 25 degrees of freedom.

Stem	Leaves	No.	Depth
0*	33	2	2
0•	667789	6	8
1*	0011112244	10	
1•	7889	4	8
2*	24	2	4
2•	9	1	2
3*		0	1
3•		0	1
4*		0	1
4•		0	1
5*	1	1	1

The stem-and-leaf display indicates that the data appears to come from a single peaked, roughly symmetric distribution whose tails die fairly rapidly except there is one outlier. We should feel reasonably comfortable with our analysis of these data. There is a possible outlier at 511.

d   To get the largest sample variance, we would add the observation of 500 since it is the farthest from the sample mean. To get the smallest sample variance, we use 142.6538 because it is the sample mean and thus will not contribute to the numerator of the formula for the sample variance.

3.73   n = 200 standby safety pumps
p = .16 = probability that a randomly selected pump failed to run after starting.
μ = np = 200 (.16) = 32
σ^2 = npq = 200(.16)(.84) = 26.88
Y = of pumps failing to run after starting

a   Pr(Y > 20) = Pr(Y* ≥ 20 + 0.5) = Pr(Y* ≥ 20.5)
Pr(Y > 20) = Pr[(Y* − np)/$\sqrt{npq}$ ≥ (20.5 − 32)/$\sqrt{26.88}$] = Pr(Z ≥ −2.22) = 1 − Pr(Z < − 2.22)
Pr(Y > 20) = 1 − .0132 = .9868

b   Pr(Y ≤ 32) = Pr(Y* ≤ 32 + .5) = Pr(Y* ≤ 32.5)
Pr(Y ≤ 32) = Pr[(Y* − np)/$\sqrt{npq}$ ≤ (32.5 − 32)/$\sqrt{26.88}$] = Pr(Z ≤ .10) = .5398

3.75   **a**   n = 600 people in the marketing study
p = .5 = probability of an individual preferring the latest model.
μ = np = 600 (.5) = 300
σ^2 = npq = 600(.5)(.5) = 150
Y = number of people preferring the latest model.

Pr(Y > 550) = Pr(Y* ≥ 550 + .5) = Pr(Y* ≥ 550.5)
Pr(Y > 550) = Pr[(Y* − np)/$\sqrt{npq}$ ≥ (550.5 − 300)/$\sqrt{150}$] = Pr(Z ≥ 20.45) ≈ 0

b n = 600; p = .25

  μ = np = 600 (.25) = 150

  σ² = npq = 600 (.25)(.75) = 112.5

  Pr(Y < 135) = Pr(Y* ≤ 135 − .5) = Pr(Y* ≤ 134.5)

  Pr(Y < 135) = Pr[(Y* − np)/$\sqrt{npq}$ ≤ (134.5 − 150)/$\sqrt{112.5}$] = Pr(Z ≤ −1.46) = .0721

3.77 n = 30

  p = probability of an exceedence = .12

  q = probability of no exceedence = .88

  Y = number of exceedences

  μ = np = (30)(.12) = 3.6

  σ² = npq = (30)(.12)(.88) = 3.168

  σ = 1.779

a Pr(Y > 10) = Pr(Y* > 10.5) = Pr[(y* − μ)/σ > (10.5 − 3.6)/1.779] = Pr(Z > 3.88) = 0

b Pr(Y ≤ 5) = Pr(Y* < 5.5) = Pr[(y* − μ)/σ < (5.5 − 3.6)/1.779] = Pr(Z > 1.07) = .8577

c Pr(2 < Y < 5) = Pr(2.5 < Y* < 4.5) = Pr[(y* − μ)/σ < (4.5 − 3.6)/1.779] = Pr[(y* − μ)/σ

    < (2.5 − 3.6)/1.779] = Pr(Z < .51) − Pr(Z < −.62) = .6950 − .2676 = .4274

d Pr(2 < Y < 5) = Pr(Y = 3) + Pr(Y = 4) = $\binom{30}{3}$(.12)³(.88)²⁷ + $\binom{30}{4}$(.12)⁴(.88)²⁶ = .2224 + .2047 = .4271

e The answers are very close, 30 is a large enough sample in this case.

3.79 a

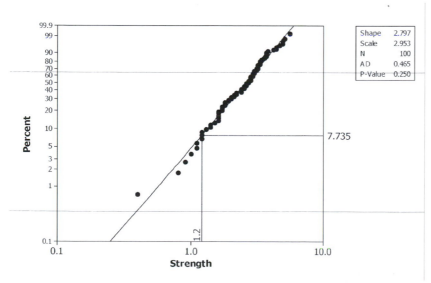

b The Weibull distribution predicts about 7.7% failures at a strength of 1.2 so their goal is not met.

3.81 a

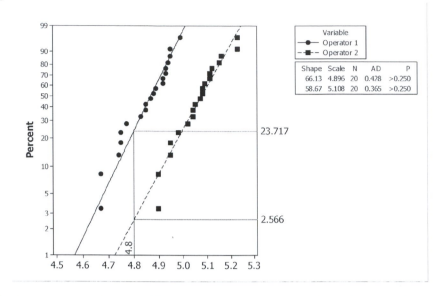

b Operator 1 has a time of 23.72, Operator 2 has a time of 2.57. The fuses made by Operator 2 last longer.

3.83 a

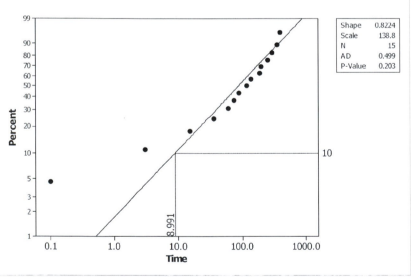

b 8.99

c $F(t) = 1 - \exp(-t/138.8)^{.8224} = .10$
$\quad\quad\exp(-t/138.8)^{.8224} = .90$
$\quad\quad(-t/138.8)^{.8224} = -.10536$
$\quad\quad t = 8.99$

3.85

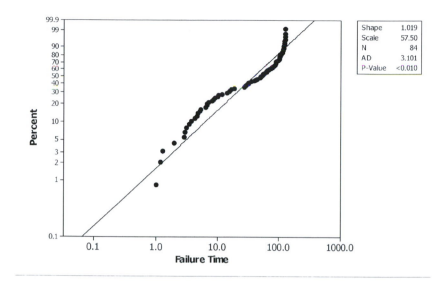

Shape	1.019
Scale	57.50
N	84
AD	3.101
P-Value	<0.010

The failure times do not seem to follow a Weibull distribution. There may be more than one type of failure mode.

CHAPTER 4

ESTIMATION AND TESTING

4.1 $\sigma = .7$ $n = 3$

    a $\bar{y} = 5.8$

       $[\bar{y} \pm z_{\alpha/2}\, \sigma / \sqrt{n}\,]$ where $\alpha = .05$ $\alpha/2 = .025$ $z_{.025} = 1.96$

       $[5.8 \pm 1.96\ (.7)/\sqrt{3}\,] = [5.8 \pm .79] = [5.01, 6.59]$ is a 95% C.I. for the true mean concentricity.

    b $z_{\alpha/2}\, \sigma / \sqrt{n} \leq B$

       $1.96\ (.7)/\sqrt{n} \leq .2$

       $\sqrt{n} \geq 1.96 \times .7/.2$

       $n \geq (1.96 \times .7/.2)^2 = 47.06$

       we must use $n = 48$

    c We assume a sample of 3 measurements is large enough to use the Central Limit Theorem.

       The baseline data can provide a basis for generating either a stem-and-leaf display or a normal probability plot. These plots can provide insight as to an appropriate minimum sample size in order to assume the Central Limit Theorem.

4.3   **a** $\sigma = .06$ $\bar{y} = 1.134$ $\alpha/2 = .025$

       $n = 4$ $\alpha = .05$

       $Z_{\alpha/2} = 1.96\ [\bar{y} \pm Z_{\alpha/2}\, \sigma / \sqrt{n}\,)] = [1.134 \pm 1.96\ (.06)/\sqrt{4}\,] = [1.134 \pm .0588]$

       So $[1.0752, 1.1928]$ is a 95% C.I. for the true mean thickness.

    b $z_{\alpha/2}\, \sigma / \sqrt{n} \leq B$

       $1.96 \times .06/\sqrt{n} \} \leq .01$

       $1.96 \times .06/.01 \geq \sqrt{n}$

       $n \geq (1.96 \times .06/.01)^2$

       $n \geq 138.2976$

       we must use $n = 139$

    c We assume a sample of size 4 thicknesses is large enough to use the Central Limit Theorem.

       The baseline data can provide a basis for generating either a stem-and-leaf diagram or a normal probability plot. These plots can provide insight as to whether a sample of 4 thicknesses is large enough to assume the Central Limit Theorem.

4.5 $\sigma = .01$ $\alpha = .01$ $\alpha/2 = .005$
$\bar{y} = \Sigma\, y/n = 7.26/8 = .9075$ $z_{.005} = 2.575$
$n = 8$

    a $[\bar{y} \pm z_{\alpha/2}\, \sigma / \sqrt{n}\,] \rightarrow [.9075 \pm 2.575\,(.01/\sqrt{8}\,)] \rightarrow [.9075 \pm .0091]$
       So [.8984, .9166] is a 99% confidence interval for the true mean yield

    b $z_{\alpha/2}\, \sigma/\sqrt{n} \rightarrow 2.575 \times .01/\sqrt{n} \le .005 \rightarrow n \ge (2.575 \times .01/.005)^2 \rightarrow n \ge 26.5225$
       we must use $n = 27$

    c   We assume a sample of size 8 yields is large enough to use the Central Limit Theorem.

       The data are single-peaked, relatively symmetric, with rapidly dying tails. A sample of 8 yields appears large enough to use the Central Limit Theorem.

4.7 $\sigma = 10$ $\alpha = .01$ $n = 10$ $\alpha/2 = .005$
$\bar{y} = \Sigma\, y/n = 669/10 = 66.9$
$z_{\alpha/2} = z_{.005} = 2.58$

    a $[\bar{y} \pm z_{\alpha/2}\, \sigma/\sqrt{n}\,] = 66.9 \pm 2.58\,(10/\sqrt{10}\,)] = [66.9 \pm 8.16]$
       So [58.75, 75.05] is a 99% C.I. for the true mean time between eruptions.

    b $B \ge z_{\alpha/2}\, \sigma/\sqrt{n} \rightarrow 5 \ge 2.58\,(10/\sqrt{n}\,)$
       $\sqrt{n} \ge (2.58)(10)/5$
       $n \ge [(2.58)(10)/5]^2 = 26.63$
       we must use $n = 27$

    c   We assume a sample of size 10 is large enough to use the Central Limit Theorem.
       The data can provide a basis for generating either a stem-and-leaf display or a normal probability plot.

4.9 $\sigma = .25$, $\alpha = .05$ $n = 7$ $\alpha/2 = .025$
$\bar{y} = \Sigma\, y/n = 16.3/7 = 2.33$
$z_{\alpha/2} = z_{.025} = 1.96$

    a $\bar{y} \pm z_{\alpha/2}\, \sigma/\sqrt{n}\,] = 2.33 \pm 1.96\,(.25/\sqrt{7}\,)] = [2.33 \pm .185]$
       So [2.14, 2.51] is a 95% C.I. for the true mean attenuation values.

    b $B \ge z_{\alpha/2}\, \sigma/\sqrt{n} \rightarrow 5 \ge 1.96\,(.25/\sqrt{n}\,)$
       $\sqrt{n} \ge (1.96)(.25)/.2$
       $n \ge [(1.96)(.25)/.2]^2 = 6.003$
       we must use $n = 7$

    c   We assume a sample of size 7 is large enough to use the Central Limit Theorem.
       The data can provide a basis for generating either a stem-and-leaf display or a normal probability plot.

4.11 $\sigma = 8$ $\mu_0 = 100$ $n = 4$

    a $\bar{y} = 101.4$

        (1) H_0: $\mu = 100$
              H_a: $\mu > 100$ $\mu = $ true mean width

        (2) $Z = \dfrac{\bar{y} - \mu}{\sigma / \sqrt{n}} = \dfrac{101.4 - 100}{8 / \sqrt{4}} = .35$

        (3) $\alpha = .01$ $z_{.01} = 2.33$
              reject H_0 if $z > 2.33$

        (4) $.35 < 2.33$, hence H_0 is not rejected.

        (5)   At $\alpha = .01$, there is insufficient evidence to reject the null hypothesis that the true mean width is 100.

    b   p-value $= Pr(Z > .35) = 1 - Pr(Z < .35) = 1 - .6368 = .3632$

    c   [91.08, 111.72] is a 99% confidence interval for μ.
       Since 100 is in the interval, we cannot reject the hypothesis that the true mean width is 100. We are 99%
       confidence that the true mean width is between 91.08 and 111.72. We reach the same conclusion
       conducting the hypothesis in part a. However, note that this hypothesis test is for a one-sided situation.

    d   Power $= Pr(\text{reject } H_0 \,|\, H_1 \text{ is true}) = Pr(Z > 2.33 \,|\, \mu = 102)$

$$= Pr\left(\frac{\bar{y} - \mu_0}{\sigma / \sqrt{n}} > 2.33 \,|\, \mu = 102\right)$$
$$= Pr(\bar{y} - \mu_0 > 2.33\ \sigma/\sqrt{n} \,|\, \mu = 102)$$
$$= Pr(\bar{y} > \mu_0 + 2.33\ \sigma/\sqrt{n} \,|\, \mu = 102)$$
$$= Pr(\bar{y} > 100 + 2.33\ (8)/\sqrt{4} \,|\, \mu = 102)$$
$$= Pr(\bar{y} > 109.32 \,|\, \mu = 102)$$
$$= Pr\left(\frac{\bar{y} - \mu}{\sigma / \sqrt{n}} > \frac{109.32 - 102}{8 / \sqrt{4}}\right)$$
$$= Pr(Z > 1.83)$$
$$= 1 - .9664$$
$$= .0336$$

    e $n \geq \left(\dfrac{(z_\alpha - z_{p1})\sigma}{\mu_1 - \mu_0}\right)^2$

       $n \geq \left(\dfrac{(2.33 - (-0.84))8}{102 - 100}\right)^2$

       $n \geq 160.7824$
       $n^* \geq 161$
       Use sample size $= 161$

    f   We have assumed that a sample size of 4 is sufficiently large for the Central Limit Theorem to apply. We
       can use a stem-and-leaf and/or a normal probability plot to check the underlying distribution of width.

4.13 $\sigma = 250 \ \mu_0 = 6500 \ n = 16$

a $\bar{y} = 6490$

(1) $H_0: \mu = 6500$
$H_a: \mu \neq 6500 \quad \mu = \text{true mean MOR}$

(2) $Z = \dfrac{\bar{y} - \mu}{\sigma / \sqrt{n}} = \dfrac{6490 - 6500}{250 / \sqrt{16}} = -.16$

(3) $\alpha = .10 \quad \alpha/2 = .05 \quad z_{.05} = 1.645$
reject H_0 if $Z > 1.645$ or $z < -1.645$

(4) $-.16 > -1.645$ and $-.16 < 1.645$, hence H_0 is not rejected.

(5) At $\alpha = .10$, there is insufficient evidence to conclude that the true mean MOR has changed from 6500.

b p-value $= 2\Pr(Z < -.16) = 2(.4364) = .8728$

c [6387.2, 6592.8] is a 90% confidence interval for μ
Since 6500 is in the interval, we cannot reject the hypothesis that the true mean MOR is 6500. We are 90% confidence that the true mean MOR is between 6387.2 and 6592.8. We reach the same conclusion conducting the hypothesis test in part a.

d Power $= \Pr(\text{reject } H_0 \mid H_1 \text{ is true}) = \Pr(Z < -1.645 \text{ or } Z > 1.645 \mid \mu = 6400)$

$= \Pr\left(\dfrac{\bar{y} - \mu_0}{\sigma / \sqrt{n}} < -1.645 \mid \mu = 6400\right) + \Pr\left(\dfrac{\bar{y} - \mu_0}{\sigma / \sqrt{n}} > 1.645 \mid \mu = 6400\right)$

$= \Pr(\bar{y} - \mu_0 < -1.645 \ \sigma/\sqrt{n} \mid \mu = 6400) + \Pr(\bar{y} - \mu_0 > 1.645 \ \sigma/\sqrt{n} \mid \mu = 6400)$

$= \Pr(\bar{y} < \mu_0 - 1.645 \ \sigma/\sqrt{n} \mid \mu = 6400) + \Pr(\bar{y} > \mu_0 + 1.645 \ \sigma/\sqrt{n} \mid \mu = 6400)$

$= \Pr(\bar{y} < 6500 - 1.645(250)/\sqrt{16} \mid \mu=6400) + \Pr(\bar{y} > 6500 + 1.645(250)/\sqrt{16} \mid \mu=6400)$

$= \Pr(\bar{y} < 6397.2 \mid \mu = 6400) + \Pr(\bar{y} > 6602.8 \mid \mu = 6400)$

$= \Pr\left(\dfrac{\bar{y} - \mu}{\sigma / \sqrt{n}} < \dfrac{6397.2 - 6400}{250 / \sqrt{16}}\right) + \Pr\left(\dfrac{\bar{y} - \mu}{\sigma / \sqrt{n}} > \dfrac{6602.8 - 6400}{250 / \sqrt{16}}\right)$

$= \Pr(Z < -.04) + \Pr(Z > 3.24)$

$= .4840 + 0$

$= .4840$

e $n \geq \left(\dfrac{(z_{\alpha/2} - z_{p1})\sigma}{\mu_1 - \mu_0}\right)^2$

$n \geq \left(\dfrac{(1.645 - (-1.04))250}{6400 - 6500}\right)^2$

$n \geq 45.1$
$n = 46$

f We have assumed that sample size of 16 is sufficiently large for the Central Limit Theorem to apply. We can use a stem-and-leaf and/or a normal probability plot to check the underlying distribution of MOR.

4.15 $\sigma = 10$ $\quad \mu_0 = 60$ $\quad$ n = 10 $\quad \bar{y} = 66.9$

a **(1)** $H_0 : \mu = 60$

$H_a : \mu > 60$ $\quad$ μ is the true mean time between eruptions

(2) $Z = \dfrac{\bar{y} - \mu}{\sigma / \sqrt{n}} = \dfrac{66.9 - 66}{10 / \sqrt{10}} = 2.18$

(3) $\alpha = .01$ $\qquad z_{.01} = 2.33$ $\qquad$ Reject H_0 if Z > 2.33

(4) 2.18 is not > 2.33, hence H_0 is not rejected

(5) At $\alpha = .01$, there is insufficient evidence to reject the null hypothesis that the true mean time between eruptions is 60

b p-value = $P(Z > 2.18) = .0146$

c [58.75,75.05] is a 99% confidence interval for the true mean time between eruptions.
Since 60 is in the interval, we cannot reject the null hypothesis that the true mean time between eruptions is 60. We are 99% confident that the true mean time between eruptions is between 58.75 and 75.05. We reach the same conclusion as part a. However, note that this hypothesis test is for a one-sided situation.

d Power $= \text{Pr}(\text{reject } H_0 \mid H_a \text{ is true}) = P(Z > 2.33 \mid \mu = 65)$

$= P\left(\dfrac{\bar{y} - \mu_0}{\sigma / \sqrt{n}} > 2.33 \mid \mu = 65 \right)$

$= P\left(\bar{y} > \mu_0 + \dfrac{2.33\sigma}{\sqrt{n}} \mid \mu = 65 \right)$

$= P\left(\bar{y} > 60 + \dfrac{2.33(10)}{\sqrt{10}} \mid \mu = 65 \right)$

$= P(\bar{y} > 67.37 \mid \mu = 65)$

$= P\left(\dfrac{\bar{y} - \mu}{\sigma / \sqrt{n}} > \dfrac{67.37 - 65}{10 / \sqrt{10}} \right)$

$= P(Z > .75)$

$= .2266$

e $n \geq \left[\dfrac{(Z_\alpha - Z_{p1})\sigma}{\mu_1 - \mu_0} \right]^2 = \left[\dfrac{(2.33 - (-1.645))10}{65 - 60} \right]^2 = 63.20$

n=64

f We have assumed that the sample size of 10 is sufficiently large for the Central Limit Theorem. We can use a stem-and-leaf displays or a normal probability plots to check the assumption.

4.17 $\sigma = .25$ $\quad \mu_0 = 2.3$ $\quad$ n = 7 $\quad \bar{y} = 2.33$

a **(1)** $H_0 : \mu = 2.3$

$H_a : \mu \neq 2.3$ $\quad$ μ is the true mean attenuation values

(2) $Z = \dfrac{\bar{y} - \mu}{\sigma / \sqrt{n}} = \dfrac{2.33 - 2.3}{.25 / \sqrt{7}} = .32$

(3) $\alpha = .05$ $\qquad z_{.025} = 1.96$ $\qquad$ Reject H_0 if Z > 1.96

(4) .32 is not > 1.96, hence H_0 is not rejected

(5) At $\alpha = .05$, there is insufficient evidence to reject the null hypothesis that the true mean attenuation values is 2.3

b   p-value = $2 \times P(|Z| > .32) = .7490$

c   [2.14, 2.52] is a 95% confidence interval for the true mean attenuation values.
Since 2.3 is in the interval, we cannot reject the null hypothesis that the true mean attenuation values is 2.3. We are 95% confident that the true mean attenuation values is between 2.14 and 2.52. We reach the same conclusion as part a.

d   Power   = Pr(reject H_0 | H_a is true = Pr($Z < -1.96$ | $\mu = 2.1$) + Pr($Z > 1.96$ | $\mu = 2.1$)

$$= P\left(\frac{\bar{y} - \mu_0}{\sigma / \sqrt{n}} < -1.96 \,\middle|\, \mu = 2.1\right) + P\left(\frac{\bar{y} - \mu_0}{\sigma / \sqrt{n}} > 1.96 \,\middle|\, \mu = 2.1\right)$$

$$= P\left(\bar{y} < \mu_0 - \frac{1.28\sigma}{\sqrt{n}} \,\middle|\, \mu = 2.1\right) + P\left(\bar{y} > \mu_0 + \frac{1.96\sigma}{\sqrt{n}} \,\middle|\, \mu = 2.1\right)$$

$$= P\left(\bar{y} < 2.3 - \frac{1.96(.25)}{\sqrt{7}} \,\middle|\, \mu = 2.1\right) + P\left(\bar{y} > 2.3 + \frac{1.96(.25)}{\sqrt{7}} \,\middle|\, \mu = 2.1\right)$$

$$= P(\bar{y} < 2.11 \,|\, \mu = 2.1) + P(\bar{y} > 2.49 \,|\, \mu = 2.1)$$

$$= P\left(\frac{\bar{y} - \mu}{\sigma / \sqrt{n}} < \frac{2.11 - 2.1}{.25 / \sqrt{7}}\right) + P\left(\frac{\bar{y} - \mu}{\sigma / \sqrt{n}} > \frac{2.49 - 2.1}{.25 / \sqrt{7}}\right)$$

$$= \text{Pr}(Z < .11) + \text{Pr}(Z > 4.13)$$

$$= .5438 + 0$$

$$= .5438$$

e $n \geq \left[\dfrac{(Z_\alpha - Z_{p1})\sigma}{\mu_1 - \mu_0}\right]^2 = \left[\dfrac{(1.96 - (-1.28)).25}{2.1 - 2.3}\right]^2 = 16.4$

n=17

f   We have assumed that the sample size of 7 is sufficiently large for the Central Limit Theorem. We can use stem-and-leaf displays or normal probability plots to check the assumption.

4.19 H_0: $\mu = \mu_0$; H_a: $\mu > \mu_0$; test statistic is $Z = \dfrac{\bar{y} - \mu_0}{\sigma / \sqrt{n}}$. We fail to reject H_0 if $Z \leq z_\alpha$

$\dfrac{\bar{y} - \mu_0}{\sigma / \sqrt{n}} \leq z_\alpha$

$\bar{y} - \mu_0 \leq z_\alpha\, \sigma / \sqrt{n}$

$-\mu_0 \leq -\bar{y} + z_\alpha\, \sigma / \sqrt{n}$

$\mu_0 \geq \bar{y} - z_\alpha\, \sigma / \sqrt{n}$

One sided confidence for μ is $(\bar{y} - Z_\alpha\, \sigma / \sqrt{n}\, , \infty)$

This interval is essentially providing a lower bound for the true mean. The advantage is that for the α value, the interval will always match the decision made from the one-sided hypothesis test. The disadvantage is that it is hard to understand and interpret the value of ∞.

4.21 a (1) $H_0: \mu = 8$

$H_a: \mu \neq 8$ μ = true mean thickness of metal wires

(2) test-statistic: $t = \dfrac{\bar{y} - \mu_0}{s / \sqrt{n}}$

(3) $\alpha = .05$ $\alpha/2 = .025$ 49 degrees of freedom
Reject if $|t| > 2.01 = t_{49,.025}$

(4) $\Sigma y = 398.8$ $\Sigma y^2 = 3182.8$ $n = 50$
$\bar{y} = \Sigma y/n = 7.976$
$s^2 = (\Sigma y^2 - (\Sigma y)^2/n)/n - 1 = (3182.8 - (398.8)^2/50)/49 = .04023$
$s = .2006$
$t = \dfrac{7.976 - 8}{.2006 / \sqrt{50}} = -.85$

(5) $|-.85| < 2.01$, so we cannot reject H_0. At $\alpha = .05$, there is insufficient evidence to conclude that the mean thickness of metal wires differs from 8 microns.

b $\left[\bar{y} \pm t_{\alpha/2,n-1} \, s/\sqrt{n} \right] = \left[7.976 \pm 2.01 (.2006/\sqrt{50}) \right] = [7.976 \pm .057]$
So [7.919, 8.033] is a 95% confidence interval for the one mean thickness of metal wires.

c $\left[\bar{y} \pm t_{\alpha/2,n-1} \, s \sqrt{1 + 1/n} \right] = \left[7.976 \pm 2.01 (.2006 \sqrt{51/50}) \right] = [7.976 \pm .407]$
So [7.57, 8.38] is a 95% prediction interval for the thicknesses.

d A stem-and-leaf diagram and/or normal probability plot can be constructed.

Stem Leaves	No.	Depth
7.6 0	1	1
7.7 00000	5	6
7.8 000000000	9	15
7.9 000000000	9	24
8.0 000000000000	12	
8.1 00	2	14
8.2 000000	6	12
8.3 0000	4	6
8.4 00	2	2

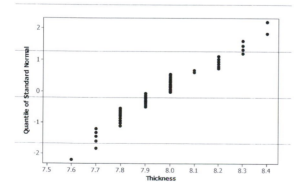

To do the analysis, we assume that the distribution of metal wire thicknesses is roughly normal (i.e., single peaked, approximately mound-shaped, tails die rapidly). The stem-and-leaf diagram demonstrates that these assumptions, while suspicious, are not grossly violated. Normal probability plot shows roughly a straight line.

e The chip-manufacturing process appears to be producing metal wires with a target thickness of 8 microns at the 95% confidence level.

4.23 a (1) $H_0: \mu = 4.5$
$H_a: \mu \neq 4.5$ μ = the mean nickel concentration

(2) test-statistic: $t = \dfrac{\bar{y} - \mu_0}{s/\sqrt{n}}$

(3) $\alpha = .05$ $\alpha/2 = .025$ $5 - 1 = 4$ df
reject if $|t| > t_{4,.025} = 2.777$

(4) $\Sigma y = 22.3$ $\Sigma y^2 = 99.65$ $n = 5$
$\bar{y} = \Sigma y/n = 22.3/5 = 4.46$
$s^2 = \dfrac{n\Sigma y^2 - (\Sigma y)^2}{n(n-1)} = \dfrac{5(99.65) - (22.3)^2}{5(4)} = .48$
$s = .2191$
$t = \dfrac{4.46 - 4.5}{.2191/\sqrt{5}} = -.408$

(5) $|-.408| < 2.777$, so we cannot reject H_0 at $\alpha = .05$. There is insufficient evidence to conclude that the mean nickel concentration has changed from 4.5oz/gal.

b $\left[\bar{y} \pm t_{\alpha/2,n-1}\, s/\sqrt{n}\right] = \left[4.46 \pm 2.777\,(.2191/\sqrt{5})\right] = [4.46 \pm .2721]$
So [4.1879, 4.7321] is a 95% confidence interval for the true mean nickel concentration.

c $\left[\bar{y} \pm t_{\alpha/2,n-1}\, s\sqrt{1 + 1/n}\right] = \left[4.46 \pm 2.777\,(.2191\sqrt{6/5})\right] = [4.46 \pm .6665]$
So [3.7935, 5.1265] is a 95% prediction interval for the nickel concentration of a day.

d To do the analysis, we assume that the distribution of the nickel concentration is roughly normal (i.e. single peaked, approximately mound-shaped, tails die rapidly).

With only 5 observations there is no good way to check these assumptions.

e The chrome-plating process appears to be able to maintain the operating standards of 4.5oz/gal. at the 95% confidence level.

4.25 a (1) $H_0: \mu = 200$
$H_a: \mu \neq 200$ μ = the true mean coating weights

(2) test-statistic: $t = \dfrac{\bar{y} - \mu_0}{s/\sqrt{n}}$

(3) $\alpha = .01$ $\alpha/2 = .005$ $30 - 1 = 29$ df
reject if $|t| > t_{29, .005} = 2.756$

(4) $\Sigma y = 6202$ $\Sigma y^2 = 1{,}283{,}330$ $n = 30$
$\bar{y} = \Sigma y/n = 6202/30 = 206.73333$
$s^2 = \dfrac{n\Sigma y^2 - (\Sigma y)^2}{n(n-1)} = \dfrac{30(1283330) - (6202)^2}{30(29)} = 40.3402$
$s = 6.3514$
$t = \dfrac{206.7333 - 200}{6.3514 / \sqrt{30}} = 5.8066$

(5) $5.8066 > 2.756$, so we reject H_0 at $\alpha = .01$. There is enough evidence to conclude that the true mean coating weight is different than 200.

b $\left[\bar{y} \pm t_{\alpha/2, n-1} \, s/\sqrt{n} \right] = \left[206.7333 \pm 2.756 \left(6.3514/\sqrt{30} \right) \right]$
Result: [203.5374, 209.9292] is a 99% confidence interval for the true mean coating weights.

c $\left[\bar{y} \pm t_{\alpha/2, n-1} \, s \sqrt{1 + 1/n} \right] = \left[206.7333 \pm 2.756 \left(6.3514 \sqrt{31/30} \right) \right] = [206.7333 \pm 17.7938]$
So [188.9395, 224.5271] is a 99% prediction interval for the coating weights of a pipe.

d To do the analysis, we assume that the distribution of coating weights is roughly normal
(i.e. single peaked, approximately mound-shaped, tails die rapidly).
The stem-and-leaf demonstrates these assumptions are not grossly violated.
The normal probability plot also shows the data follow a straight line very well.
We should feel comfortable with the assumptions.

Stem	Leaves	No.	Depth
19*	3	1	1
19•	68	2	3
20*	0222344444	10	13
20•	56667888	8	
21*	2223	4	9
21•	56688	5	5

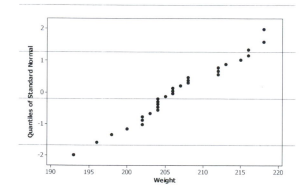

e The coating process appears to process higher coating weight than the standard of 200 pounds per pipe at the 99% confidence level.

4.27 a (1) $H_0: \mu = 100$

$H_a: \mu \neq 100$ μ = true mean particle sizes in microns.

(2) test-statistic: $t = \dfrac{\overline{y} - \mu_0}{s / \sqrt{n}}$

(3) $\alpha = .05$ $\alpha/2 = .025$ $25 - 1 = 24$ df
reject for $|t| > t_{24,.025} = 2.064$

(4) $\Sigma y = 2487.7$ $\Sigma y^2 = 247874.19$ $n = 25$

$\overline{y} = \Sigma y / n = 2487.7/25 = 99.508$

$s^2 = \dfrac{n\Sigma y^2 - (\Sigma y)^2}{n(n-1)} = \dfrac{25(247874.19) - (2487.7)^2}{25(24)} = 13.6724$

$s = 3.6976$

$t = \dfrac{99.508 - 100}{3.6976 / \sqrt{25}} = -.6653$

(5) $|{-.6653}| < 2.064$, so we cannot reject H_0. There is insufficient evidence to conclude that the true mean particle sizes is not 100 microns.

b $\left[\overline{y} \pm t_{\alpha/2, n-1} \, s/\sqrt{n} \right] = \left[99.508 \pm 2.064 \left(3.6976/\sqrt{25} \right) \right] = [99.508 \pm 1.52637]$
So [97.9816, 101.0344] is a 95% confidence interval for the true mean particle size.

c $\left[\overline{y} \pm t_{\alpha/2, n-1} \, s \sqrt{1 + 1/n} \right] = \left[99.508 \pm 2.064 \left(3.6976 \sqrt{26/25} \right) \right] = [99.508 \pm 7.7830]$
So [91.7250, 107.2910] is a 95% prediction interval for the particle size for a future day.

d To do the analysis, we assume that the distribution of particles sizes is roughly normal (i.e. single peaked, approximately mound-shaped, tails die rapidly). The stem-and-leaf demonstrates these assumptions may be violated since it is slightly skewed and double peaked. However, the normal probability plot shows the data follows a straight line quite well. We should feel comfortable with the assumptions.

Stem Leaves	No.	Depth
9t 23	2	2
9f 445	3	5
9s 66677	5	10
9• 89	2	12
10* 00011	5	
10t 222223	6	8
10f 45	2	2

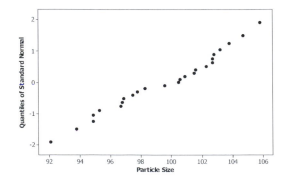

e This process appears to have met the target specification of 100 microns well at 95% confidence level.

4.29 a (1) $H_0 : \mu = 196$

$H_a : \mu \neq 196$ μ is the true mean resistivity

(2) test-statistic: $t = \dfrac{\overline{y} - \mu_0}{s / \sqrt{n}}$

(3) $\alpha = .10$ d.f. $= n-1 = 7$

$t_{7,.10} = 1.895$, Reject H_0 if $t > 1.895$

(4) $\sum y = 1568.9$ $\sum y^2 = 307680.14$

$\overline{y} = 1568.9 / 8 = 196.112$

$s^2 = (307680.14 - (1568.9)^2 / 8) / 7 = .0204$

$s = \sqrt{.0204} = .143$

$t = \dfrac{196.112 - 196}{.143 / \sqrt{8}} = 2.22$

since $2.22 > 1.895$, we reject H_0

(5) At $\alpha = .10$, there is sufficient evidence to conclude that the true mean resistivity has increased.

b $\left[\overline{y} \pm t_{\alpha/2, n-1}(s / \sqrt{n}) \right] = \left[196.112 \pm 1.895(.143 / \sqrt{8}) \right] = [196.112 \pm .096] = [196.016, 196.208]$

is a 90% confidence interval for the true mean resistivity.

c $\left[\overline{y} \pm t_{\alpha/2, n-1} s(\sqrt{1 + (1/n)}) \right] = \left[196.112 \pm 1.895(.143)\sqrt{\dfrac{9}{8}} \right] = [196.112 \pm .287] = [195.825, 196.399]$

is a 90% prediction interval for the resistivities.

d To do the analysis, we assume that the distribution of resistivities is roughly normal. With only 8 observations, there is no good way to check assumptions. The normal probability plot shows that data follows roughly a straight line. Since there are only 8 observations, a slight departure from a straight line is acceptable. We should feel comfortable with our assumptions.

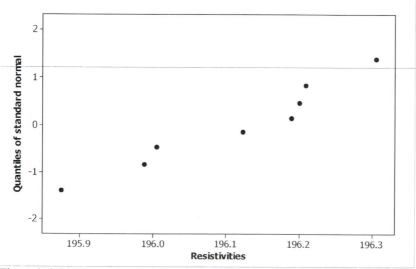

e The resistivities are not at their nominal value at the 95% confidence level.

4.31 a (1) $H_0 : p = .12$

$H_a : p < .12$ p is the true proportion of landings in exceedence

(2) $Z = \dfrac{\hat{p} - p_0}{\sqrt{p_0 q_0 / n}}$ is the test-statistic, where $p_0 = .12$, $q_0 = .88$, $n = 250$

Since $np_0 = 30$, $nq_0 = 220$, we can use the approximation.

(3) $\alpha = .05$ $Z_{.05} = -1.645$, Reject H_0 if $Z < -1.645$

(4) $\hat{p} = \dfrac{22}{250} = .088$ $Z = \dfrac{.088 - .12}{\sqrt{.12(.88) / 250}} = -1.56$

since -1.56 is not less than -1.645, we fail to reject H_0

(5) At $\alpha = .05$, there is insufficient evidence to conclude that the true proportion in exceedence has been lowered

b $\left[\hat{p} \pm Z_{\alpha/2} \sqrt{\tfrac{\hat{p}\hat{q}}{n}} \right] = \left[.088 \pm 1.96 \sqrt{\dfrac{.088(.812)}{250}} \right] = [.088 \pm .033] = [.053, .123]$

is a 95% confidence interval for the true proportion in exceedence

4.33 a (1) H_0: $p = .10$

H_a: $p > .10$ p = true proportion of bores outside the specifications

(2) $Z = \dfrac{\hat{p} - p_0}{\sqrt{p_0 q_0 / n}}$ is the test statistic

$p_0 = .10$ $q_0 = 1 - .10 = .90$ $n = 165$ $np_0 = 16.5$ $nq_0 = 148.5$

(3) $\alpha = .01$ $z_{.01} = 2.33$,   Reject H_0 if $Z > 2.33$

(4) $\hat{p} = 36/165 = .218$ $Z = \dfrac{.218 - .10}{\sqrt{(.10)(.90) / 165}} = 5.05$

Since $5.06 > 2.33$ we reject H_0

(5) At $\alpha = .01$, there is sufficient evidence to conclude that the proportion of bores outside of the specifications exceeds 10%.

b $\left[\hat{p} \pm Z_{\alpha/2} \sqrt{\hat{p}\hat{q} / n} \right] = \left[.218 \pm 2.57 \sqrt{(.218)(.782) / 165} \right] = [.218 \pm .0826] = [.1354, 3006]$

is a 99% confidence interval for the true proportion of bores outside the specifications.

4.35 a (1) $H_0: p = .10$

$H_a: p > .10$ p = true proportion of resistors with resistance outside the specifications.

(2) $Z = \dfrac{\hat{p} - p_0}{\sqrt{p_0 q_0 / n}}$ is the test statistic

$p_0 = .10$ $q_0 = .90$ $n = 180$ $np_0 = 18$ $nq_0 = 162$

(3) $\alpha = .05$ $z_{.05} = 1.645$; reject H_0 if test statistic $Z > 1.645$

(4) $\hat{p} = 46/180 = .26$ $Z = \dfrac{.26 - .10}{\sqrt{(.10)(.90) / 180}} = 7.16$

(5) Since $7.16 > 1.645$, we reject H_0. At $\alpha = .05$, there is sufficient evidence to conclude that the proportion of resistors with resistance outside the specifications exceeds 10%.

b $\left[\hat{p} \pm Z_{\alpha/2} \sqrt{\hat{p}\hat{q} / n} \right] = \left[.26 \pm 1.96 \sqrt{(.26)(.74) / 180} \right] = [.26 \pm .064] = [.196, .324]$

95% confidence interval for the true proportion of resistors outside the specifications.

4.37 a (1) $H_0: p = .07$

$H_a: p < .07$ (to test for an improvement)

p = true proportion of drive trains failing to meet specification.

(2) $Z = \dfrac{\hat{p} - p_0}{\sqrt{p_0 q_0 / n}}$ is the test statistic

$n = 200$ $p_0 = .07$ $q_0 = .93$ $np_0 = 14$ $nq_0 = 186$

(3) $\alpha = .05$ $z_{.05} = 1.645$; reject H_0 if test statistic $Z < -1.645$.

(4) $\hat{p} = 12/200 = .06$ $Z = \dfrac{.06 - .07}{\sqrt{(.07)(.93) / 200}} = -.55$

(5) $-.55 > -1.645$, so we cannot reject H_0. At $\alpha = .05$, there is insufficient evidence to conclude that the true proportion of drive trains failing to meet specifications has decreased.

b $\alpha = .05$ $\alpha/2 = .025$ $z_{.025} = 1.96$

$\left[\hat{p} \pm Z_{\alpha/2} \sqrt{\hat{p}\hat{q} / n} \right] = \left[.06 \pm 1.96 \sqrt{(.06)(.94) / 200} \right] = [.06 \pm .03] = [.03, .09]$

is a 95% confidence interval for the true proportion of drive trains which fail.

4.39 a

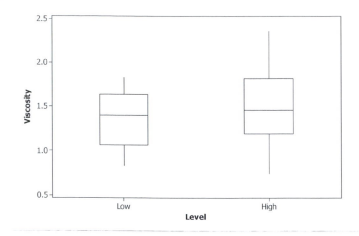

b (1) $H_0: \mu_H - \mu_L = 0$
$H_a: \mu_H - \mu_L > 0$
μ_H = true average thickness for high viscosity
μ_L = true average thickness for low viscosity

(2) test-statistic: $t = \dfrac{\bar{y}_H - \bar{y}_L - 0}{s_p\sqrt{\dfrac{1}{n_H} + \dfrac{1}{n_L}}}$

(3) $\alpha = .05$ $n_1 + n_2 - 2 = 16 + 16 - 2 = 30$ df
$t_{.05,30} = 1.697$; reject H_0 if $t > 1.697$

(4) $\Sigma y_L = 21.57$ $n = 16$ $\bar{y}_L = 1.348$
$\Sigma y_L^2 = 30.7981$ $s_L^2 = [30.7981 - (21.57)^2/16]/15 = .1146$

$\Sigma y_H = 23.79$ $n = 16$ $\bar{y}_H = 1.487$
$\Sigma y_H^2 = 39.0843$ $s_H^2 = [39.0843 - (23.79)^2/16]/15 = .2474$

$s_p^2 = [(n_1 - 1)s_1^2 + (n_2 - 1)s_2^2]/(n_1 + n_2 - 2) = [15(.1146) + 15(.2474)]/30 = .181$
$s_p = .4254$

$t = \dfrac{\bar{y}_H - \bar{y}_L - 0}{s_p\sqrt{\dfrac{1}{n_H} + \dfrac{1}{n_L}}} = \dfrac{1.487 - 1.348}{.4254\sqrt{\dfrac{1}{16} + \dfrac{1}{16}}} = .924$

(5) $.924 < 1.697$, so we cannot reject H_0. At $\alpha = .05$, there is insufficient evidence to conclude that the average thickness for high viscosity exceeds the average thickness for low viscosity.

c $\alpha = .05$ $\alpha/2 = .025$ $n_1 + n_2 - 2 = 16 + 16 - 2 = 30$ df $t_{.025,30} = 2.042$

$$\left[(\bar{y}_H - \bar{y}_L) \pm t_{.025,30}\, s_p\sqrt{\dfrac{1}{n_H} + \dfrac{1}{n_L}}\right] = \left[(1.487 - 1.348) \pm 2.042(.4254)\sqrt{\dfrac{1}{16} + \dfrac{1}{16}}\right]$$

which is $[.139 \pm .307] = [-.168, .446]$, a 95% confidence interval for the true difference in the mean coating thickness for high-low viscosity.

d The two samples were independent of each other. The two populations of viscosities have a common variance. The distributions of thicknesses of both the high and low viscosities are both reasonably normal or mound-shaped. There is some concern about the normality in the thickness from the high viscosity.

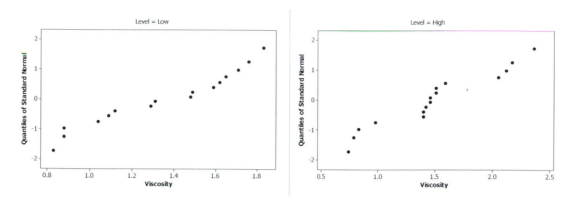

e Fail to reject H_0 at the $\alpha = .05$ level. There is insufficient evidence to conclude the higher viscosity leads to thicker coatings; appears to agree with the boxplots.

4.41 a

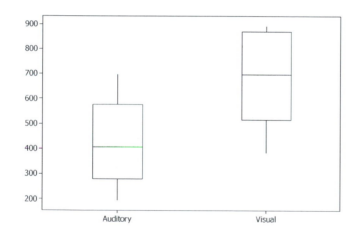

b (1) H_0: $\mu_a - \mu_v = 0$
H_a: $\mu_a - \mu_v \neq 0$
μ_a = true average response due to auditory stimuli
μ_v = true average response due to visual stimuli

(2) test-statistic: $t = \dfrac{\bar{y}_a - \bar{y}_v - 0}{s_p \sqrt{\dfrac{1}{n_a} + \dfrac{1}{n_v}}}$

(3) $\alpha = .01$ $\alpha/2 = .005$ $n_a + n_v - 2 = 18$ df
$t_{18,.005} = 2.879$; reject H_0 if $|t| > 2.879$

(4) $n_a = 10$ $\bar{y}_a = 422.7$ $s_a^2 = 31106.68$
      $n_v = 10$ $\bar{y}_v = 677.7$ $s_v^2 = 39587.79$
      $s_p = 188.01$

$$t = \frac{\bar{y}_a - \bar{y}_v - 0}{s_p\sqrt{\dfrac{1}{n_a} + \dfrac{1}{n_v}}} = \frac{422.7 - 677.7}{188.01\sqrt{\dfrac{1}{10} + \dfrac{1}{10}}} = -3.03$$

(5) Since $|-3.03| > 2.879$, we reject H_0. At $\alpha = .01$, there is sufficient evidence to conclude a difference exists between the average responses of auditory and visual stimuli.

c $\left[(\bar{y}_a - \bar{y}_v) \pm t_{.025,18}\, s_p\sqrt{\dfrac{1}{n_a} + \dfrac{1}{n_v}} \right] = \left[(422.7 - 677.7) \pm 2.879(188.01)\sqrt{\dfrac{1}{10} + \dfrac{1}{10}} \right]$

which is $[-255 \pm 242.07] = [-497.07, -12.93]$, a 99% confidence interval for the time difference between the average response due to auditory stimuli and the average response due to visual stimuli.

d The distribution of the number of wrist movements due to aural stimuli and the distribution of the number of wrist movements due to visual stimuli are both approximately normal or mound-shaped. The samples are independent of each other. The two distributions share a common variance. One can be comfortable with these assumptions, but may wish to try a transformation because of how the normal probability plots look in the tails.

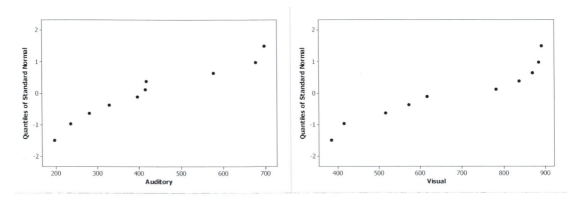

e Reject H_0 at $\alpha = .01$ level. There is sufficient evidence to conclude a difference between auditory and visual mean wrist movements. The boxplots confirm that the visual mean is larger. The confidence interval suggests between 13 and 497 movements, which is a rather large interval.

4.43 a

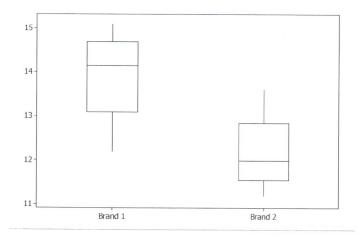

b (1) $H_0: \mu_1 - \mu_2 = 0$
$H_a: \mu_1 - \mu_2 \neq 0$
μ_1 = average maximum output in fluid ounces/hr. due to brand 1
μ_2 = average maximum output in fluid ounces/hr. due to brand 2

(2) test-statistic: $t = \dfrac{\overline{y}_1 - \overline{y}_2 - 0}{s_p \sqrt{\dfrac{1}{n_1} + \dfrac{1}{n_2}}}$

(3) $\alpha = .10$ $\alpha/2 = .05$ $n_1 + n_2 - 2 = 4 + 4 - 2 = 6$ df
$t_{.05,6} = 1.943$; reject H_0 if $|t| > 1.943$

(4) $n_1 = 4$ $\overline{y}_1 = 13.9$ $s_1^2 = 1.5$
$n_2 = 4$ $\overline{y}_2 = 12.2$ $s_2^2 = 1.02$
$s_p = \sqrt{\dfrac{3(1.5) + 3(1.02)}{6}} = 1.122$

$t = \dfrac{\overline{y}_1 - \overline{y}_2 - 0}{s_p \sqrt{\dfrac{1}{n_1} + \dfrac{1}{n_2}}} = \dfrac{13.9 - 12.2}{1.122 \sqrt{\dfrac{1}{4} + \dfrac{1}{4}}} = 2.14$

(5) $|2.14| > 1.943$, we reject H_0. At $\alpha = .10$, there is sufficient evidence to reject the hypothesis that the average outputs of the brands are the same.

c $\left[(\overline{y}_1 - \overline{y}_2) \pm t_{.05,6}\, s_p \sqrt{\dfrac{1}{n_1} + \dfrac{1}{n_2}} \right] = \left[(13.9 - 12.2) \pm 1.943(1.122)\sqrt{\dfrac{1}{4} + \dfrac{1}{4}} \right]$
which is [1.7 ± 1.54] = [.16, 3.24], a 90% confidence interval for $\mu_1 - \mu_2$.

d The distributions of outputs are approximately normal for both brand 1 and brand 2. The distributions share a common variance. The samples from each distribution are independent of each other. With only samples of size 4, it is hard to assess assumptions.

e Reject H_0 at the alpha = .10 level. Sufficient evidence to conclude a difference between brand 1 and brand 2 outputs. In fact, brand 1 larger agreeing with the boxplots and confidence interval. Note in the boxplots the third quartile for brand 2 is smaller than the first quartile of brand 1.

4.45 a

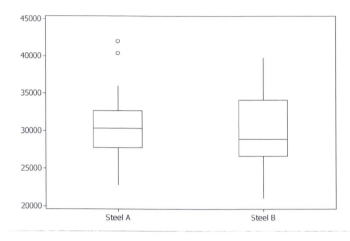

b (1) $H_0: \mu_A - \mu_B = 0$
$H_a: \mu_A - \mu_B \neq 0$
μ_A = true mean fracture load for steel A
μ_B = true mean fracture load for steel B

(2) $t = \dfrac{\bar{y}_A - \bar{y}_B - 0}{s_p\sqrt{\dfrac{1}{n_A} + \dfrac{1}{n_B}}}$ is the test statistic

(3) $\alpha = .05$ $n_A + n_B - 2 = 30 + 30 - 2 = 58$ df
$t_{.025,58} = 2.001$ Reject H_0 if $|t| > 2.001$

(4) $\sum y_A = 910500$ $\sum y_A^2 = 28245210000$ $n_A = 30$
$\sum y_B = 904530$ $\sum y_B^2 = 28012347500$ $n_B = 30$
$\bar{y}_A = 910500/30 = 30350$ $\bar{y}_B = 904530/30 = 30151$
$s_A^2 = (28245210000 - (910500)^2/30)/29 = 21087414$
$s_B^2 = (28012347500 - (904530)^2/30)/29 = 25512533$
$s_p^2 = \dfrac{(n_A - 1)s_A^2 + (n_B - 1)s_B^2}{n_A + n_B - 2} = \dfrac{21087414 + 25512533}{2} = 23299973.5$
$s_p = \sqrt{23299973.5} = 4827.005$
$t = \dfrac{30350 - 30151}{4827.005\sqrt{\dfrac{2}{30}}} = .16$

(5) Since $|.16|$ is not greater than 2.001, we fail to reject H_0. At $\alpha = .05$, there is sufficient evidence to conclude that the true mean fracture loads are equal

c $\left[\overline{y}_A - \overline{y}_B \pm t_{.025,58}s_p\sqrt{\dfrac{1}{n_A}+\dfrac{1}{n_B}}\right]=\left[(30350-30151)\pm 2.001(4827.005)\sqrt{2/30}\right]$

$=[199\pm 2493.9]=[-2294.9,2692.9]$ is a 95% confidence interval for the true difference between the mean fracture load for steel A and that of steel B

d   We assume that the two steel samples are independent of each other. The two populations from Steel A and Steel B have common variance. The distribution of fracture loads for both steels are reasonably normal.

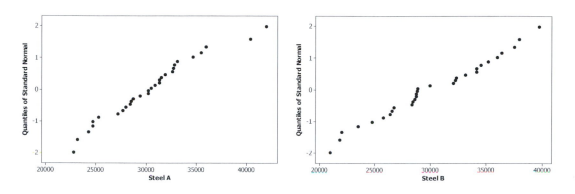

Since both probability plots follow roughly a straight line, the assumptions seem to be satisfied.

e   Fail to reject H_0 at α=.05. Insufficient evidence to conclude there is a difference between the two steel types.

4.47 a

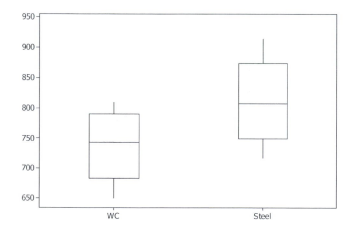

b   (1) H_0: $\mu_A - \mu_B = 0$
H_a: $\mu_A - \mu_B \neq 0$
μ_A = true mean peak temperature for WC
μ_B = true mean peak temperature for Steel

(2) $t = \dfrac{\bar{y}_A - \bar{y}_B - 0}{s_p\sqrt{\dfrac{1}{n_A} + \dfrac{1}{n_B}}}$ is the test statistic

(3) $\alpha = .10$ $n_A + n_B - 2 = 4 + 4 - 2 = 6$ df
$t_{6,.05} = 1.943$ Reject H_0 if $|t| > 1.943$

(4) $\sum y_A = 2948$ $\sum y_A^2 = 2187130$ $n_A = 4$
$\sum y_B = 3247$ $\sum y_B^2 = 2656565$ $n_B = 4$
$\bar{y}_A = 2948/4 = 737$ $\bar{y}_B = 3247/4 = 811.8$
$s_A^2 = (2187130 - (2948)^2/4)/3 = 4818$
$s_B^2 = (2656565 - (3247)^2/4)/3 = 6937.6$
$s_p^2 = \dfrac{(n_A - 1)s_A^2 + (n_B - 1)s_B^2}{n_A + n_B - 2} = \dfrac{4818 + 6937.6}{6} = 5877.8$
$s_p = \sqrt{5877.8} = 76.67$
$t = \dfrac{737 - 811.8}{76.67\sqrt{\dfrac{2}{4}}} = -1.38$

(5) Since $|-1.38|$ is not greater than 1.943, we fail to reject H_0. At $\alpha = .10$, there is insufficient evidence to conclude that the true mean peak temperatures differ

c $\left[\bar{y}_A - \bar{y}_B \pm t_{.05,6}s_p\sqrt{\dfrac{1}{n_A} + \dfrac{1}{n_B}}\right] = \left[(737 - 811.8) \pm 1.943(76.67)\sqrt{2/4}\right]$
$= [-74.8 \pm 105.34] = [-180.14, 30.54]$ is a 90% confidence interval for the true difference between the mean peak temperature for WC and that of steel

d We assume that the two temperature samples are independent of each other. The two populations of WC and steel have common variance. The distribution of peak temperature for both holders are reasonably normal. There is no good way to check the normality assumption with a sample of just size four.

e Fail to reject H_0 at $\alpha = .10$. There is insufficient evidence to conclude there is a difference between the two holders.

4.49 a (1) H_0: $\delta = 0$
H_a: $\delta \neq 0$ δ difference between average length measurement for Graham and Brian

(2) $t = \dfrac{\bar{d}}{s_d/\sqrt{n}}$ is the test-statistic

(3) $\alpha = .05$ $n - 1 = 13$ df
$t_{13,.025} = 2.160$ Reject H_0 if $|t| > 2.160$

(4)

String	Difference	String	Difference
1	−.2	8	−.4
2	−.1	9	.1
3	.2	10	−.1
4	−.4	11	−.2
5	1.2	12	.7
6	.3	13	.3
7	−.6	14	−.6

$\sum d = .2 \quad \bar{d} = .2/14 = .0143 \quad \sum d^2 = 3.3$

$s_d^2 = \dfrac{(3.3 - (.2)^2/14)}{13} = .254 \qquad s_d = \sqrt{.254} = .504$

$t = \dfrac{.0143}{.504/\sqrt{14}} = .11$

(5) Since |.11| is not greater than 2.160, we fail to reject H_0. At $\alpha = .05$, there is insufficient evidence to conclude that the true means differ for the two raters.

b $\left[\bar{d} \pm t_{.025,13}(s_d/\sqrt{n}) \right] = \left[.0143 \pm 2.16(.504/\sqrt{14}) \right] = [.0143 \pm .291] = [−.277, .3053]$

is a 95% confidence interval for the true difference between mean lengths for Graham and Brian

c We assume that the distribution of differences in lengths is roughly normal. We should feel comfortable with the assumptions.

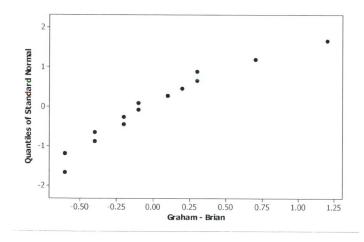

4.51 a (1) $H_0: \delta = 0$

$H_a: \delta \neq 0 \qquad \delta$ difference between average running time for the two operators

(2) $t = \dfrac{\bar{d}}{s_d/\sqrt{n}}$ is the test-statistic

(3) $\alpha = .05 \quad \alpha/2 = .025 \qquad n - 1 = 20 - 1 = 19$ df

$t_{19,.025} = 2.093 \qquad$ Reject H_0 if $|t| > 2.093$

(4)

Fuse	Difference	Fuse	Difference
1	−.24	11	−.21
2	−.11	12	−.20
3	−.20	13	−.21
4	−.25	14	−.24
5	−.23	15	−.19
6	−.18	16	−.28
7	−.23	17	−.21
8	−.21	18	−.18
9	−.23	19	−.19
10	−.23	20	−.19

$$\bar{d} = -.2105 \quad s_d = .0349 \qquad t = \frac{-.2105}{0.349 / \sqrt{20}} = -26.97$$

(5) $|-26.97| > 2.093$, hence reject H_0. At $\alpha = .05$, there is sufficient evidence to indicate that a difference exists between the average running time of operator 1 and the average running time of operator 2.

b $\left[\bar{d} \pm t_{19,.025}\, s_d/\sqrt{n}\right] = \left[-.2105 \pm 2.093(.0349/\sqrt{20}\,)\right] = [-.2105 \pm .0163] = [-.2268, -.1942]$ is a 95% confidence interval for the difference between the average running time for Operator 1 and the average running time for Operator 2.

c We assume that the distribution of differences is roughly normal (i.e. single peaked rapidly dying tails). A stem-and-leaf diagram of the differences is below

```
−2* 58
−2• 001111333344
−1* 88999
−1• 1
```

We should feel comfortable with our assumptions as the stem-and-leaf is single reached with most of the data falling in the low-20s. The tails die rapidly.

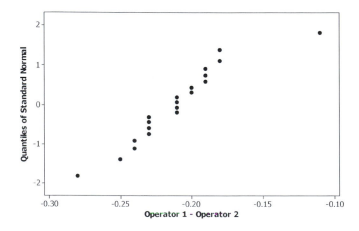

4.53 a (1) H_0: $\delta = 0$

H_a: $\delta > 0$ δ = difference between average thickness of plates made at location A and
average thickness of plates made at location B

(2) $t = \dfrac{\overline{d}}{s_d / \sqrt{n}}$ is the test-statistic

(3) $\alpha = .05$ $n - 1 = 10 - 1 = 9$ df
$t_{9,.05} = 1.833$ Reject H_0 if $t > 1.833$

(4)

Plate	Difference	Plate	Difference
1	1.35	6	0.00
2	1.35	7	−0.70
3	0.75	8	0.10
4	0.00	9	0.70
5	2.00	10	0.15

$\overline{d} = 5.7/10 = .57$ $s_d = .8145$ $t = \dfrac{.57}{.8145 / \sqrt{10}} = 2.213$

(5) Because 2.213 > 1.833, we reject H_0. At $\alpha = .05$, there is sufficient evidence to conclude that the
average thickness of plates made at location A exceed the average thickness of plates made at location
B.

b $\alpha = .05$ $\alpha/2 = .025$ $n - 1 = 9$ df $t_{9,.025} = 2.262$
$\left[\overline{d} \pm t_{9,.025}\, s_d/\sqrt{n}\,\right] = \left[.57 \pm 2.262(.8145/\sqrt{10}\,)\right] = [.57 \pm .58] = [-.01, 1.15]$ is a 95% confidence interval for
the true difference in the average thickness of plates made at location A and the average thickness of plates
made at location B. Note that the hypothesis test was a one sided procedure but this confidence interval is
two-sided in nature.

c We assume that the distribution of differences in plate thicknesses is roughly normal (i.e., single peaked,
rapidly dying tails). Looking at normal probability plot of differences, it appears we can be comfortable
with these assumptions.

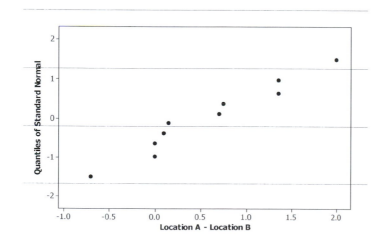

4.55 a (1) $H_0: \delta = 0$

$H_a: \delta \neq 0$ δ = difference between the average temperature of instrument 1 and instrument 2

(2) $t = \dfrac{\overline{d}}{s_d / \sqrt{n}}$ is the test-statistic

(3) $\alpha = .05$ $\alpha/2 = .025$ $n-1 = 5-1 = 4$ df

$t_{4,.025} = 2.777$ Reject H_0 if $|t| > 2.777$

(4)

Day	1	2	3	4	5
Difference	$-.6$	$-.6$	~~$-.5$~~	$-.7$	$-.7$

$\Sigma d = -3.1$ $\Sigma d^2 = 1.95$ $\overline{d} = -3.1/5 = -.62$

$s_d^2 = (1.95 - (-3.1)^2/5)/4 = .007$ $s_d = .0837$

$t = \dfrac{-.62}{.0837 / \sqrt{5}} = -16.56$

(5) $|-16.56| > 2.777$, reject H_0. At $\alpha = .05$, there is sufficient evidence to conclude that a difference exists between the average temperatures read off instruments 1 and instrument 2.

b $\left[\overline{d} \pm t_{4,.025}\, s_d/\sqrt{n} \right] = [-.62 \pm 2.777(.0839/\sqrt{5})] = [-.62 \pm .10] = [-.72, -.52]$ is a 95% confidence interval for the true mean difference between temperature readings from instrument 1 and instrument 2. Instrument 2 gives larger readings on average.

c We assume that the distribution of differences of temperature readings in roughly normal (i.e., single peaked, tails die rapidly), but with a sample of size 5 it is difficult to assess assumptions.

4.57 a (1) $H_0: p_1 - p_2 = 0$

$H_a: p_1 - p_2 \neq 0$ p_1 = true proportion in exceedence for airline 1

p_2 = true proportion in exceedence for airline 2

(2) $Z = \dfrac{(\hat{p}_1 - \hat{p}_2) - 0}{\sqrt{\hat{p}\hat{q}\left(\dfrac{1}{n_1} + \dfrac{1}{n_2}\right)}}$ is the test statistic

$\hat{p}_1 = \dfrac{14}{156} = .0897$ $\hat{p}_2 = \dfrac{11}{198} = .0556$ $\hat{p} = \dfrac{14+11}{156+198} = .0706$ $\hat{q} = 1 - \hat{p} = .9294$

(3) $\alpha = .05$ $Z_{.025} = 1.96$; reject H_0 if $|Z| > 1.96$

(4) $Z = \dfrac{.0897 - .0556}{\sqrt{(.0706)(.9294)\left(\dfrac{1}{156} + \dfrac{1}{198}\right)}} = 1.24$

(5) $1.24 < 1.96$, so we cannot reject H_0. There is insufficient evidence to conclude that the true proportion in exceedence is different for the two airlines with $\alpha = .05$.

b $\left[(\hat{p}_1 - \hat{p}_2) \pm Z_{\alpha/2} \sqrt{ \left(\dfrac{\hat{p}_1 \hat{q}_1}{n_1} + \dfrac{\hat{p}_2 \hat{q}_2}{n_2} \right) } \right] = \left[.0341 \pm 1.96 \sqrt{ \dfrac{.0897(.9103)}{156} + \dfrac{.0556(.9444)}{198} } \right] =$

$[.0341 \pm .055] = [-021, .089]$
is a 95% confidence interval for the true difference between the proportion of exceedence for airlines 1 and 2.

4.59 a (1) H_0: $p_1 - p_2 = 0$
$\quad\quad\quad$ H_a: $p_1 - p_2 \neq 0$ $\quad\quad$ p_1 = true proportion of failures with heterotaxy syndrome
$\quad\quad\quad\quad\quad\quad\quad\quad\quad\quad\quad\quad$ p_2 = true proportion of failures without heterotaxy syndrome

(2) $Z = \dfrac{(\hat{p}_1 - \hat{p}_2) - 0}{\sqrt{\hat{p}\hat{q}\left(\dfrac{1}{n_1} + \dfrac{1}{n_2} \right)}}$ is the test statistic

$\hat{p}_1 = \dfrac{9}{41} = .219 \quad \hat{p}_2 = \dfrac{75}{459} = .163 \quad \hat{p} = \dfrac{9 + 75}{41 + 459} = .168 \quad \hat{q} = 1 - \hat{p} = .832$

(3) $\alpha = .01 \quad Z_{.005} = 2.58$; reject H_0 if $|Z| > 2.58$

(4) $Z = \dfrac{.219 - .163}{\sqrt{(.168)(.832)\left(\dfrac{1}{41} + \dfrac{1}{459} \right)}} = .92$

(5) $.92 < 2.58$, so we cannot reject H_0. There is insufficient evidence to conclude that the true proportion of failures is different with or without heterotaxy syndrome using $\alpha = .01$.

b $\left[(\hat{p}_1 - \hat{p}_2) \pm Z_{\alpha/2} \sqrt{ \left(\dfrac{\hat{p}_1 \hat{q}_1}{n_1} + \dfrac{\hat{p}_2 \hat{q}_2}{n_2} \right) } \right] = \left[.056 \pm 2.58 \sqrt{ \dfrac{.219(.781)}{41} + \dfrac{.163(.837)}{459} } \right] =$

$[.056 \pm .167] = [-.111, .223]$
is a 99% confidence interval for the true difference between the proportion of failures with heterotaxy syndrome and without heterotaxy syndrome.

4.61 a (1) H_0: $p_1 - p_2 = 0$
$\quad\quad\quad$ H_a: $p_1 - p_2 \neq 0$ $\quad\quad$ p_1 = true proportion to meet specifications from operator
$\quad\quad\quad\quad\quad\quad\quad\quad\quad\quad\quad\quad$ p_2 = true proportion to meet specifications from trainer

(2) $Z = \dfrac{(\hat{p}_1 - \hat{p}_2) - 0}{\sqrt{\hat{p}\hat{q}\left(\dfrac{1}{n_1} + \dfrac{1}{n_2} \right)}}$ is the test statistic

$\hat{p}_1 = \dfrac{38}{190} = .2 \quad \hat{p}_2 = \dfrac{50}{195} = .256 \quad \hat{p} = \dfrac{38 + 50}{190 + 195} = .229 \quad \hat{q} = 1 - \hat{p} = .771$

(3) $\alpha = .10 \quad Z_{.05} = 1.645$; reject H_0 if $|Z| > 1.645$

(4) $Z = \dfrac{.2 - .256}{\sqrt{(.229)(.771)\left(\dfrac{1}{190} + \dfrac{1}{195}\right)}} = -1.31$

(5) $|-1.31| < 1.645$, so we cannot reject H_0. There is insufficient evidence to conclude that the true proportion meeting specification is different for the operator and the trainer, $\alpha = .10$.

b $\left[(\hat{p}_1 - \hat{p}_2) \pm Z_{\alpha/2} \sqrt{\left(\dfrac{\hat{p}_1 \hat{q}_1}{n_1} + \dfrac{\hat{p}_2 \hat{q}_2}{n_2} \right)} \right] = \left[-.056 \pm 1.645 \sqrt{\dfrac{.2(.8)}{190} + \dfrac{.256(.744)}{195}} \right] =$

$[-.056 \pm .07] = [-.126, .014]$
is a 90% confidence interval for the true difference between the proportion meeting specification for the operator and the trainer.

4.63 a (1) H_0: $\sigma^2 = 3600$
$\quad\quad\quad\quad$ H_a: $\sigma^2 \neq 3600$ $\quad\quad\quad$ $\sigma^2 =$ population *variance* ($= 60^2$) for these concentrations

$\quad\quad$ **(2)** The test statistic is $\chi^2 = 25s^2/3600$ $\quad\quad$ (d.f. = 25)

$\quad\quad$ **(3)** Reject H_0 if $\chi^2 < \chi^2_{25,.975} = 13.120$ or $\chi^2 > \chi^2_{25,.025} = 40.646$

$\quad\quad$ **(4)** $s^2 = 9644.075$ $\quad\quad$ so $\chi^2 = 25(9644.075)/3600 = 66.9733$

$\quad\quad$ **(5)** Reject H_0 since $66.9733 > 40.646$. We have sufficient evidence to conclude that the variability is not equal to the claim of 3600.

$\quad$ **b**   A 95% confidence interval for σ^2 is $[(25)(9644.075)/40.646, (25)(9644.075)/13.120]$
$\quad\quad$ $[5931.75 , 18376.67]$

$\quad\quad$ Clearly the variance is greater than 3600 and so the standard deviation is above 60.

$\quad$ **c**   The 26 samples need to form a random sample from a normal population. One checks this by using a normal probability plot. Note that the observation of 511 appears usual.

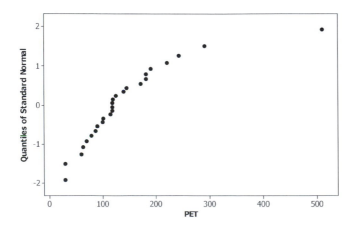

4.65 a (1) H_0: $\sigma^2 = 144$

H_a: $\sigma^2 \neq 144$ σ^2 = pop. variance of the particle sizes

(2) The test statistic is $\chi^2 = 24s^2/144$ (d.f. = 24)

(3) Reject H_0 if $\chi^2 < \chi^2_{24,.975} = 12.401$ or $\chi^2 > \chi^2_{24,.025} = 39.364$

(4) $s^2 = 13.67$ so $\chi^2 = 24(13.67)/144 = 2.278$

(5) Reject H_0 because $2.278 < 12.401$. We have sufficient evidence to conclude that the variability is in fact less than 144.

b A 95% confidence interval for σ^2 is $[24(13.67)/39.364, 24(13.67)/12.401] = [8.335, 26.456]$

c One must assume that this is a random sample from a normal population. A normal probability plot is used to check this, and it appears that one can be comfortable with this assumption.

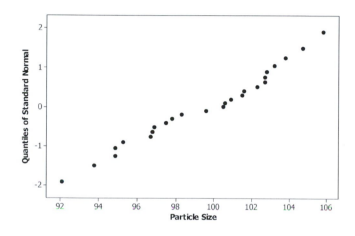

4.67 a (1) H_0: $\sigma^2 = 70$

H_a: $\sigma^2 > 70$ σ^2 = population variance of the deflections

(2) The test statistic is $\chi^2 = (n-1)s^2/\sigma^2$ $n - 1 = 9 - 1 = 8$ df

(3) $\alpha = .05$, so reject H_0 if $\chi^2 > \chi^2_{8,.05} = 15.507$

(4) $s^2 = 78.75$ so $\chi^2 = 8(78.75)/70 = 9$

(5) Since $9 < 15.507$ (not greater than), we fail to reject H_0. At $\alpha = .05$, there is insufficient evidence to conclude that the population variance is not more than 70.

b $\left[\dfrac{(n-1)s^2}{\chi^2_{8,.025}}, \dfrac{(n-1)s^2}{\chi^2_{8,.975}} \right] = \left[\dfrac{(9-1)78.75}{17.534}, \dfrac{(9-1)78.75}{2.180} \right]$

Thus $[35.93, 288.99]$ is a 95% confidence interval for the true variance of deflections.

c   We assume that we have a random sample from a normal distribution.
We should feel comfortable with the assumptions.

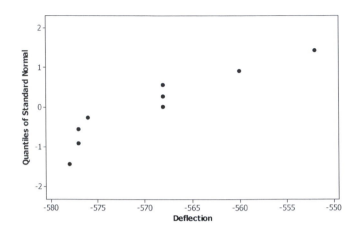

4.69 **a** $[(n-1)s^2/\chi^2_{n-1,\alpha}, \infty) = [9(.9)/16.919, \infty) = [.48, \infty)$ is a one-sided 95% confidence interval for the true population variance of the whiteness measures from the new vendor. We can conclude that the true population variance is higher than .4, since .4 is not in the interval, with 95% confidence. (Note: $\chi^2_{9,.05} = 16.919$)

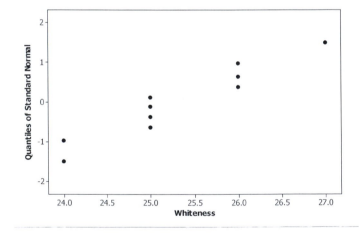

b   By checking the normal probability plot, the data forms a roughly straight line pattern. Since the sample size is small, a slight departure from a straight line is acceptable. The assumptions should not be seriously violated.

4.71 a (1) $H_0 : \sigma_1^2 = \sigma_2^2$

$H_a : \sigma_1^2 \neq \sigma_2^2$

σ_1^2 = population variance of ultimate strength for ratio 1:1

σ_2^2 = population variance of ultimate strength for ratio 1:2

(2) $F = \dfrac{\text{larger sample variance}}{\text{smaller sample variance}}$ is the test-statistic

(3) $\alpha = .10$ num d.f. = den d.f. = $4 - 1 = 3$ Reject H_0 if $F > F_{3,3,.05} = 9.28$

(4) $s_1^2 = 342676$ $s_2^2 = 716051$ $F = \dfrac{716051}{342676} = 2.09$

(5) Since 2.09 is not greater than 9.28, we fail to reject H_0. At $\alpha = .10$, there is insufficient evidence to conclude that the two population variances differ.

b We assume that the two culture samples are independent of each other and the distribution of strength for both methods are reasonably normal. There is no good way to check the normality assumption with only a sample of size four.

4.73 a (1) $H_0 : \sigma_1^2 = \sigma_2^2$

$H_a : \sigma_1^2 \neq \sigma_2^2$

σ_1^2 = population variance of errors for novice inspectors

σ_2^2 = population variance of errors for experienced inspectors

(2) $F = \dfrac{\text{larger sample variance}}{\text{smaller sample variance}}$ is the test-statistic

(3) $\alpha = .10$ num d.f. = den d.f. = $12 - 1 = 11$ Reject H_0 if $F > F_{11,11,.05} = 2.82$

(4) $s_1^2 = 96.97$ $s_2^2 = 32.99$ $F = \dfrac{96.97}{32.99} = 2.94$

(5) Since 2.94 > 2.82, we reject H_0. At $\alpha = .10$, there is sufficient evidence to conclude that the two population variances for novice and experienced inspectors differ.

b We assume that the two inspector samples are independent of each other and the distribution of errors for both methods are reasonably normal. From the plots, we should feel comfortable with our assumptions.

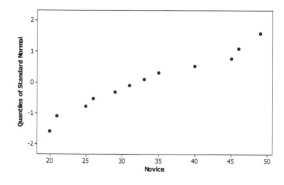

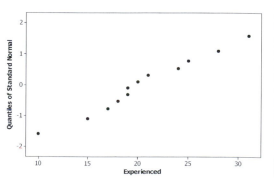

4.75 a (1) $H_0 : \sigma_1^2 = \sigma_2^2$

$H_a : \sigma_1^2 \neq \sigma_2^2$

$\sigma_1^2 =$ population variance of pollutants with traffic $= 2$

$\sigma_2^2 =$ population variance of pollutants with traffic $= 3$

(2) $F = \dfrac{\text{larger sample variance}}{\text{smaller sample variance}}$ is the test-statistic

(3) $\alpha = .10$ num d.f. $=$ den d.f. $= 6 - 1 = 5$ Reject H_0 if $F > F_{5,5,.05} = 5.05$

(4) $s_1^2 = 1.005$ $s_2^2 = .488$ $F = \dfrac{1.005}{.488} = 2.06$

(5) Since 2.06 is not greater than 5.05, we fail to reject H_0. At $\alpha = .10$, there is insufficient evidence to conclude that the two population variances for pollutants with traffic $= 2$ and traffic $= 3$ differ.

b We assume that the two samples are independent of each other and the distribution of pollutants for both methods are reasonably normal. There is no good way to check the normality assumption with only a sample of size six. Traffic $= 3$ looks fine but traffic $= 2$ is suspect.

4.77 a There are a few large points that affect the normal plot.

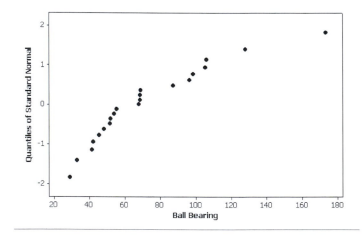

b Sign test of median = 80.00 versus < 80.00

	N	Below	Equal	Above	P	Median
Ball Bearing	21	14	0	7	0.0946	67.80

Since the p-value $= .0946 < .10$, we reject H_0.

c No, with the t-test we fail to reject H_0.

d The natural log transformation works well.

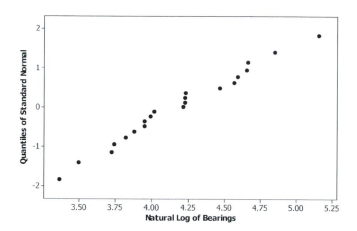

e `Test of mu = 4.382 vs < 4.382`

Variable	N	Mean	StDev	SE Mean	T	P
ln-ball	22	4.12168	0.52545	0.11203	-2.32	0.015

Reject H_0, we get the same conclusion as part b.

4.79 a   The normal plots do not show a straight line.

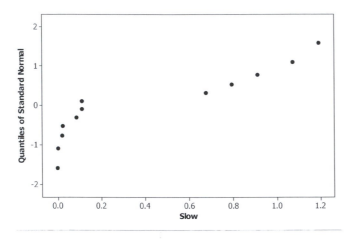

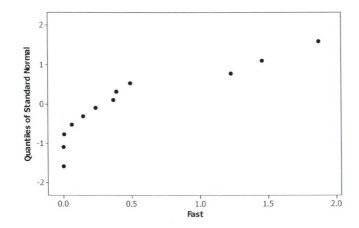

b

```
         N  Median
Slow    12  0.1114
Fast    12  0.3045

Point estimate for ETA1-ETA2 is -0.0381
90.0 Percent CI for ETA1-ETA2 is (-0.3682,0.2832)
W = 143.0
Test of ETA1 = ETA2 vs ETA1 not = ETA2. p-value = 0.7075
```

We fail to reject H₀, there insufficient evidence to conclude there is a difference in the calcium uptake for fast and slow muscles.

4.81 a There is a curved pattern in the normal plot for Stress = .87.

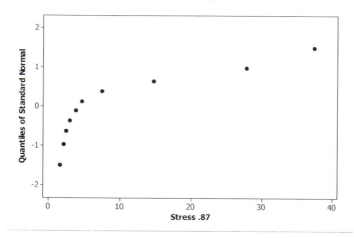

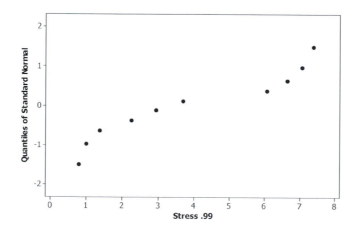

b
```
                    N   Median
Stress .87   10    4.30
Stress .99   10    3.33

Point estimate for ETA1-ETA2 is 1.49
91.1 Percent CI for ETA1-ETA2 is (-1.19,8.04)
W = 122.0
Test of ETA1 = ETA2 vs ETA1 not = ETA2, p-value = 0.2123
```

We fail to reject H_0, there insufficient evidence to conclude there is a difference in the failure times for stress = .87 and stress = .99.

4.83 a The normal plot shows roughly a straight line.

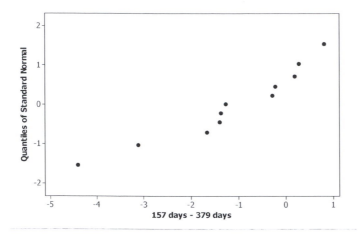

b Paired T for 157 days - 379 days not = 0

```
                   N       Mean     StDev   SE Mean
157 days          11    4.76818   2.08611   0.62899
379 days          11    5.89636   2.17329   0.65527
Difference        11   -1.12818   1.55314   0.46829

95% CI for mean difference: (-2.17160, -0.08477)
T-Value = -2.41   P-Value = 0.037
```

We reject H_0, there is sufficient evidence to conclude that the VEP levels are higher for 379 days.

c Test of median = 0 versus median not = 0

```
              N for    Wilcoxon                Estimated
         N    Test    Statistic       P         Median
diff-VEP 11    11         9.5        0.041      -0.8875
```

We reject H_0, there is sufficient evidence to conclude that the VEP levels are higher for 379 days.
This is the same conclusion as part b.

CHAPTER 5

CONTROL CHARTS AND STATISTICAL PROCESS CONTROL

5.1 $\theta_0 = 1 = \mu_0$ $\sigma = .06$ $n = 4$ $\hat{\theta} = \bar{y}$
$\text{LCL} = \theta_0 - 3\ \sigma_{\hat{\theta}} = 1 - 3(.06/\sqrt{4}) = .91$
$\text{UCL} = \theta_0 + 3\ \sigma_{\hat{\theta}} = 1 + 3\ (.06/\sqrt{4}) = 1.09$

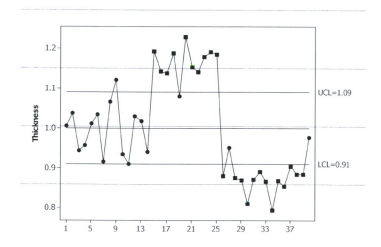

The process does not appear in control over the observation period. It goes above the control limit at sample 9, comes back within the control limits, and then goes out of control for all but one sample between 15 and 25. It appears the company made an adjustment after sample 25 but over compensated for the excess thickness causing the process to fall below the control limits in all but 2 of the remaining samples.

5.3 $\theta_0 = 100 = \mu_0$ $\qquad$ $\sigma = 8$ $\qquad$ $n = 4$ $\qquad$ $\hat{\theta} = \bar{y}$

$\qquad$ LCL $= \theta_0 - 3\,\sigma_{\hat{\theta}} = 100 - 3(8/\sqrt{4}) = 88$

$\qquad$ UCL $= \theta_0 + 3\,\sigma_{\hat{\theta}} = 100 + 3\,(8/\sqrt{4}) = 112$

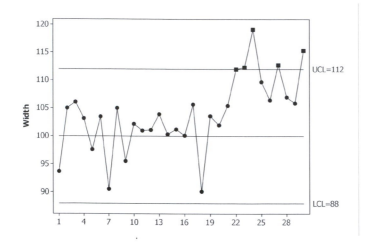

The chart signals several possible but out of control situations. A shift appears to have occurred around subgroups 21-24. The plastic injection molding process appears to be unstable and requires attention.

5.5 $\theta_0 = 10.5 = \mu_0$ $\qquad$ $\sigma = .1$ $\qquad$ $n = 5$ $\qquad$ $\hat{\theta} = \bar{y}$

$\qquad$ LCL $= \theta_0 - 3\,\sigma_{\hat{\theta}} = 10.5 - 3(.1/\sqrt{5}) = 10.366$

$\qquad$ UCL $= \theta_0 + 3\,\sigma_{\hat{\theta}} = 10.5 + 3\,(.1/\sqrt{5}) = 10.634$

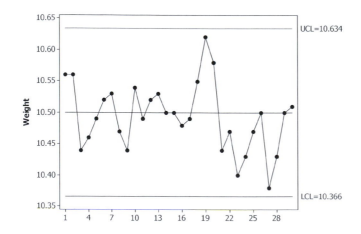

The packaging line appears to be in control. No signals for out of control.

5.7 $\theta_0 = 98 = \mu_0$ $\sigma = 4$ $n = 4$ $\hat{\theta} = \bar{y}$

$LCL = \theta_0 - 3\,\sigma_{\hat{\theta}} = 98 - 3(4/\sqrt{4}) = 92$

$UCL = \theta_0 + 3\,\sigma_{\hat{\theta}} = 98 + 3\,(4/\sqrt{4}) = 104$

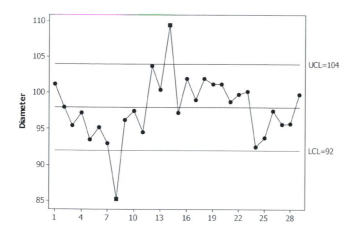

The chart indicates two out of control subgroups.

5.9 a $\bar{R} = 2.155$ $n = 5$ $m = 20$

$LCL = D_3\,\bar{R} = 0(2.155) = 0$

$UCL = D_4\,\bar{R} = 2.115(2.155) = 4.55782$

One "out of control" subgroup, process variation appears stable otherwise.

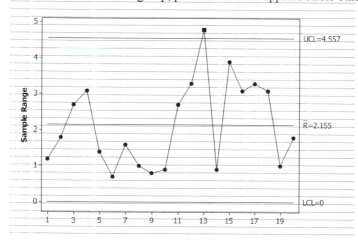

b $\bar{\bar{y}} = 2.63$ $n = 5$ $m = 20$
  $\text{LCL} = \bar{\bar{y}} - A_2 \bar{R} = 2.63 - 0.577\,(2.155) = 1.38$
  $\text{UCL} = \bar{\bar{y}} + A_2 \bar{R} = 2.63 + 0.577\,(2.155) = 3.873$

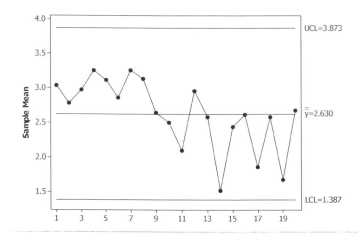

The process location appears in control. No out of control signals.

c   The R-chart assumes: 1) the process is in control during the base period, and 2) the data follow a normal distribution.

The $\bar{X}$-chart based on the sample range assumes: 1) the R-chart is in control, 2) the subgroup means are in control during the base period, and 3) the sample size is large enough to assume that the subgroup means follow a normal distribution by the Central Limit Theorem.

We should next look at a stem-and-leaf display of the data, during the base period. The following display appear roughly to follow a bell-shaped curve; reasonably comfortable with normality assumptions.

```
N = 100    Median = 2.7
Quartiles = 1.8, 3.2

Decimal point is at the colon

      1     1    0 : 4
      3     2    0 : 89
     12     9    1 : 011222344
     29    17    1 : 56666667777888899
     41    12    2 : 000112224444
           22    2 : 5555666677788888889999
     37    20    3 : 00001112222223333344
     17     9    3 : 566777778
      8     3    4 : 244
      5     3    4 : 799
      2     1    5 : 1
      1     1    5 : 6
```

5.11 a $\bar{R} = 7.2$ $n = 5$ $m = 20$

$LCL = D_3 \bar{R} = 0(7.2) = 0$

$UCL = D_4 \bar{R} = 2.115(7.2) = 15.228$

Process variation appears "in control".

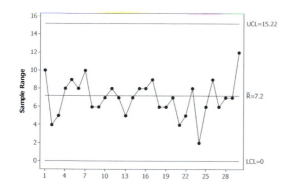

b $\bar{\bar{y}} = 34.06$ $n = 5$ $m = 20$

$LCL = \bar{\bar{y}} - A_2 \bar{R} = 34.06 - 0.577\,(7.2) = 29.91$

$UCL = \bar{\bar{y}} + A_2 \bar{R} = 34.06 + 0.577\,(7.2) = 38.21$

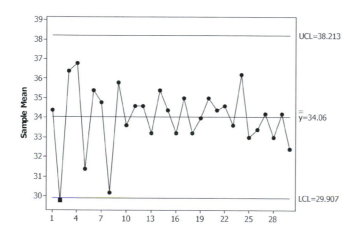

The process location appears "in control".

c The R-chart assumes: 1) the process is in control during the base period, and 2) the data follow a normal distribution.

The $\overline{X}$ -chart based on the sample range assumes: 1) the R-chart is in control, 2) the subgroup means are in control during the base period, and 3) the sample size is large enough to assume that the subgroup means follow a normal distribution by the Central Limit Theorem.

We should next look at a stem-and-leaf display of the data, during the base period. The following display indicates a process with multiple peaks uncomfortable with normality assumptions. Consider stratification.

```
N = 100   Median = 35   Quartiles = 31, 36
Decimal point is at the colon

    3     3    28 : 000
    9     6    29 : 000000
   19    10    30 : 0000000000
   28     9    31 : 000000000
   31     3    32 : 000
   36     5    33 : 00000
   48    12    34 : 000000000000
         23    35 : 00000000000000000000000
   29     7    36 : 0000000
   22     5    37 : 00000
   17    10    38 : 0000000000
    7     3    39 : 000
    4     3    40 : 000
    1     1    41 : 0
```

5.13 a $\overline{R} = 7.3$ n = 5   m = 20

$\text{LCL} = D_3 \overline{R} = 0(7.3) = 0$

$\text{UCL} = D_4 \overline{R} = 2.115(7.3) = 15.44$

Process variation appears "in control".

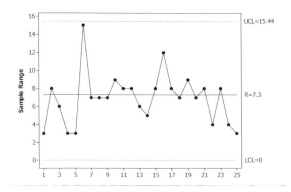

b $\bar{\bar{y}} = 19.36$ $n = 5$ $m = 20$

$\text{LCL} = \bar{\bar{y}} - A_2 \bar{R} = 19.36 - 0.577 \, (7.3) = 15.15$

$\text{UCL} = \bar{\bar{y}} + A_2 \bar{R} = 19.36 + 0.577 \, (7.3) = 23.57$

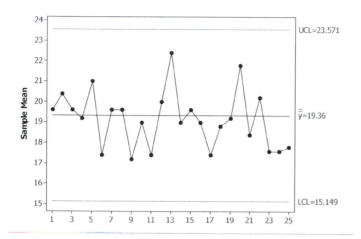

The process location appears "in control".

c The R-chart assumes: 1) the process is in control during the base period, and 2) the data follow a normal distribution.

The $\bar{X}$-chart based on the sample range assumes: 1) the R-chart is in control, 2) the subgroup means are in control during the base period, and 3) the sample size is large enough to assume that the subgroup means follow a normal distribution by the Central Limit Theorem.

We should next look at a stem-and-leaf display of the data, during the base period. Since the display indicates a bell-shaped curve, we can be comfortable with the normality assumptions.

```
N = 100    Median = 19    Quartiles = 17, 22
Decimal point is at the colon

      1      1    10 : 0
      1      0    11 :
      2      1    12 : 0
      3      1    13 : 0
      4      1    14 : 0
     10      6    15 : 000000
     17      7    16 : 0000000
     29     12    17 : 000000000000
     39     10    18 : 0000000000
            19    19 : 0000000000000000000
     42      5    20 : 00000
     37      9    21 : 000000000
     28     11    22 : 00000000000
     17      6    23 : 000000
     11      5    24 : 00000
      6      5    25 : 00000
      1      1    26 : 0
```

5.15 a $\bar{R} = .546$ $n = 3$ $m = 20$

$LCL = D_3\bar{R} = 0(.551) = 0$

$UCL = D_4\bar{R} = 2.575(.546) = 1.406$

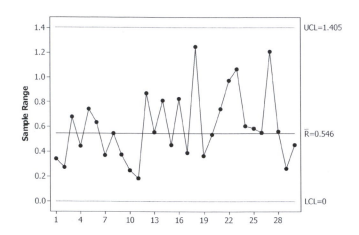

The process variation appears stable. No out-of-control signals.

b $\bar{\bar{y}} = 2.170$ $n = 3$ $m = 20$

$LCL = \bar{\bar{y}} - A_2\bar{R} = 2.170 - 1.023(.546) = 1.611$

$UCL = \bar{\bar{y}} + A_2\bar{R} = 2.170 + 1.023(.546) = 2.728$

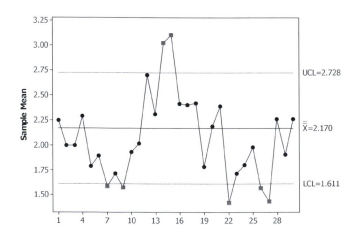

The chart indicates several out-of-control subgroups with a cyclical pattern.

c The R-chart assumes: 1) the process is in control during the base period and 2) the data follow a normal distribution

The $\bar{X}$-chart based on the sample range assumes: 1) the R-chart is in control, 2) the subgroup means are in control during the base period, and 3) the sample size is large enough to assume that the subgroup means follow a normal distribution by the Central Limit Theorem

We should next look at a stem-and-leaf display of the data during the base period. The following display appears roughly normal.

```
Stem-and-leaf   N  = 60
Leaf Unit = 0.10

Stem            Leaves
2     1         33
6     1         4555
16    1         6666666777
23    1         8899999
      2         000000001111
25    2         22222333
17    2         445555
11    2         67
9     2         88999
4     3         01
2     3         3
1     3         4
```

d $\bar{R}=1.101$ $n=9$ $m=10$

$$LCL = D_3\bar{R} = .184(1.101) = .203$$
$$UCL = D_4\bar{R} = 1.816(1.101) = 1.999$$

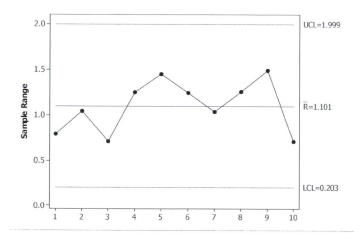

The process variation appears stable. No out-of-control signals.

e $\bar{\bar{y}} = 2.072$ $n = 9$ $m = 10$

$LCL = \bar{\bar{y}} - A_2\bar{R} = 2.072 - .337(1.101) = 1.701$

$UCL = \bar{\bar{y}} + A_2\bar{R} = 2.072 + .337(1.101) = 2.443$

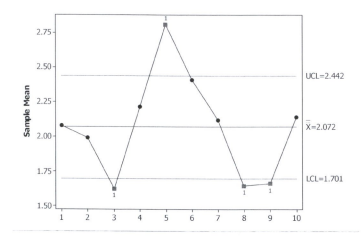

The chart indicates several out-of-control subgroups with a cyclical pattern.

f The R-chart assumes: 1) the process is in control during the base period and 2) the data follow a normal distribution

The $\bar{X}$-chart based on the sample range assumes: 1) the R-chart is in control, 2) the subgroup means are in control during the base period, and 3) the sample size is large enough to assume that the subgroup means follow a normal distribution by the Central Limit Theorem

We should next look at a stem-and-leaf display of the data during the base period. The following display appears roughly normal.

```
Stem-and-leaf    N  = 90        Leaf Unit = 0.10

Stem              Leaves
1      0          7
2      0          9
3      1          1
7      1          2333
13     1          445555
28     1          666666666777777
41     1          8889999999999
       2          0000000000001111
33     2          22222223333
21     2          444555555
12     2          677
9      2          88999
4      3          01
2      3          3
1      3          4
```

g The process variation is stable in both a and d. However, there is much more variation when using cassette as the unit. Both charts for the mean show several out-of-control points with the same cyclical pattern.

5.17 a $\bar{s}^2 = .97475$ $n = 5$ $m = 20$

$LCL = \chi^2_{4,.00135}\ \bar{s}^2/4 = .106(.97475)/4 = .026$

$UCL = \chi^2_{4,.99865}\ \bar{s}^2/4 = 17.8(.97475)/4 = 4.34$

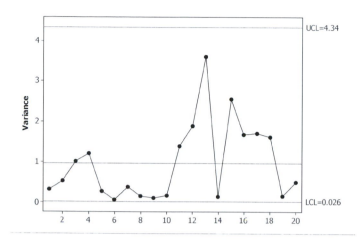

Process variation appears stable.

b $\bar{\bar{y}} = 2.63$

$LCL = \bar{\bar{y}} - 3\sqrt{\bar{s}^2/5} = 2.63 - 3\sqrt{.97474/5} = 1.305$

$LCL = \bar{\bar{y}} + 3\sqrt{\bar{s}^2/5} = 2.63 + 3\sqrt{.97474/5} = 3.955$

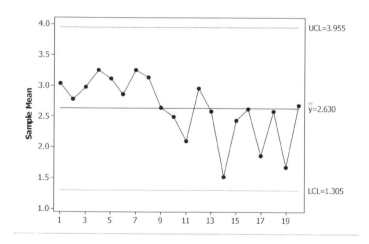

Process location appears "in control".

c The s^2-chart assumes: 1) the process is in control during the base period, and 2) the data follow a normal distribution.

The $\overline{X}$-chart based on the sample variance assumes: 1) the s^2-chart is in control, i.e. the process variance is constant, 2) the subgroup means are in control during the base period, and 3) the sample size is large enough to assume that the subgroup means follow a normal distribution by the Central Limit Theorem.

The following display appears roughly to follow a bell-shaped curve; so we are comfortable with the normality assumption.

```
N = 100    Median = 2.7
Quartiles = 1.8, 3.2

Decimal point is at the colon

      1     1    0 : 4
      3     2    0 : 89
     12     9    1 : 011222344
     29    17    1 : 56666667777888899
     41    12    2 : 000112224444
           22    2 : 5555666677788888889999
     37    20    3 : 00001112222223333344
     17     9    3 : 566777778
      8     3    4 : 244
      5     3    4 : 799
      2     1    5 : 1
      1     1    5 : 6
```

5.19 a $\overline{s}^2 = 8.465$ $n = 5$ $m = 20$

$LCL = \chi^2_{4,.00135}\, \overline{s}^2/4 = .106(8.465)/4 = .224$

$UCL = \chi^2_{4,.99865}\, \overline{s}^2/4 = 17.8(8.465)/4 = 37.67$

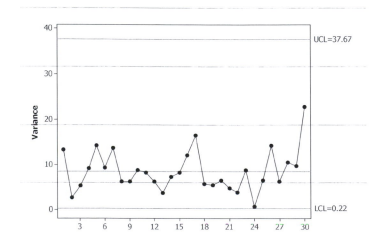

Process variation appears stable.

b $\bar{\bar{y}} = 34.06$

$$LCL = \bar{\bar{y}} - 3\sqrt{s^2/5} = 34.06 - 3\sqrt{8.465/5} = 30.16$$

$$UCL = \bar{\bar{y}} + 3\sqrt{s^2/5} = 34.06 + 3\sqrt{8.465/5} = 37.96$$

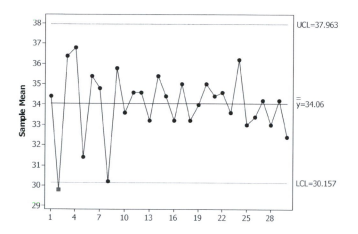

Second subgroup signals "out of control" process location appears stable otherwise.

c The s^2-chart assumes: 1) the process is in control during the base period, and 2) the data follow a normal distribution.

The $\bar{X}$-chart based on the sample variance assumes: 1) the s^2-chart is in control, i.e. the process variance is constant, 2) the subgroup means are in control during the base period, and 3) the sample size is large enough to assume that the subgroup means follow a normal distribution by the Central Limit Theorem. Indication of multiple peaks in the following display, so we are uncomfortable with the normality assumption.

```
N = 100    Median = 35    Quartiles = 31, 36

Decimal point is at the colon
    3    3    28 : 000
    9    6    29 : 000000
   19   10    30 : 0000000000
   28    9    31 : 000000000
   31    3    32 : 000
   36    5    33 : 00000
   48   12    34 : 000000000000
        23    35 : 00000000000000000000000
   29    7    36 : 0000000
   22    5    37 : 00000
   17   10    38 : 0000000000
    7    3    39 : 000
    4    3    40 : 000
    1    1    41 : 0
```

5.21 a $\bar{s}^2 = 10.025$ $n = 5$ $m = 20$

$\text{LCL} = \chi^2_{4,.00135} \, \bar{s}^2/4 = .106(10.025)/4 = .266$

$\text{UCL} = \chi^2_{4,.99865} \, \bar{s}^2/4 = 17.8(10.025)/4 = 44.61$

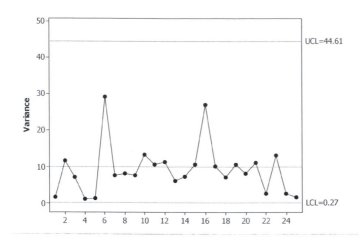

Process variation appears "in control"

b $\bar{\bar{y}} = 19.36$

$\text{LCL} = \bar{\bar{y}} - 3\sqrt{\bar{s}^2/5} = 19.36 - 3\sqrt{10.025/5}$
$= 15.11$

$\text{LCL} = \bar{\bar{y}} + 3\sqrt{\bar{s}^2/5} = 19.36 + 3\sqrt{10.025/5}$
$= 23.61$

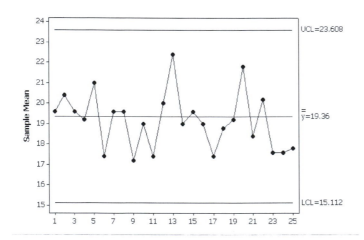

Process location clearly "in control".

c The s^2-chart assumes: 1) the process is in control during the base period, and 2) the data follow a normal distribution.

The $\overline{X}$-chart based on the sample variance assumes: 1) the s^2-chart is in control, i.e. the process variance is constant, 2) the subgroup means are in control during the base period, and 3) the sample size is large enough to assume that the subgroup means follow a normal distribution by the Central Limit Theorem. This display looks bell-shaped, so the normality assumption looks OK.

```
N = 100    Median = 19
Quartiles = 17, 22

Decimal point is at the colon
      1     1    10 : 0
      1     0    11 :
      2     1    12 : 0
      3     1    13 : 0
      4     1    14 : 0
     10     6    15 : 000000
     17     7    16 : 0000000
     29    12    17 : 000000000000
     39    10    18 : 0000000000
           19    19 : 0000000000000000000000
     42     5    20 : 00000
     37     9    21 : 000000000
     28    11    22 : 00000000000
     17     6    23 : 000000
     11     5    24 : 00000
      6     5    25 : 00000
      1     1    26 : 0
```

5.23 a $\overline{s}^2 = 0.1045$ $n = 3$ $m = 20$

$$LCL = \frac{\chi^2_{2,.00135}(\overline{s}^2)}{n-1} = \frac{.003(.1045)}{2} \approx 0$$

$$UCL = \frac{\chi^2_{2,.99865}(\overline{s}^2)}{n-1} = \frac{13.215(.1045)}{2} = .691$$

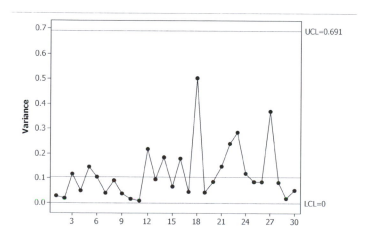

The process variation appears stable. No out-of-control signals.

b $\bar{\bar{y}} = 2.170$ $n = 3$ $m = 20$

$$LCL = \bar{\bar{y}} - 3\sqrt{\frac{s^2}{n}} = 2.170 - 3(.1866) = 1.61$$

$$UCL = \bar{\bar{y}} + 3\sqrt{\frac{s^2}{n}} = 2.170 + 3(.1866) = 2.73$$

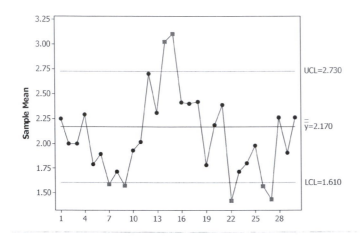

There are several out-of-control subgroups with a cyclical pattern.

c   The s^2-chart assumes: 1) the process is in control during the base period, 2) the data follow a normal distribution

The $\bar{X}$ -chart based on the sample variance assumes: 1) the s^2-chart is in control, 2) the subgroup means are in control during the base period, 3) the sample size is large enough to assume that the subgroup means follow a normal distribution by the Central Limit Theorem.

We should next look at a stem-and-leaf display of the data during the base period. The following display appears roughly normal.

```
Stem-and-leaf   N  = 60
Leaf Unit = 0.10

Stem           Leaves
2     1        33
6     1        4555
16    1        6666666777
23    1        8899999
      2        000000001111
25    2        22222333
17    2        445555
11    2        67
9     2        88999
4     3        01
2     3        3
1     3        4
```

d $\bar{s}^2 = .1439$ $n = 9$ $m = 10$

$$LCL = \frac{\chi^2_{8,.99865}(\bar{s}^2)}{n-1} = \frac{.931(.1439)}{8} = .017$$

$$UCL = \frac{\chi^2_{8,.00135}(\bar{s}^2)}{n-1} = \frac{25.361(.1439)}{8} = .456$$

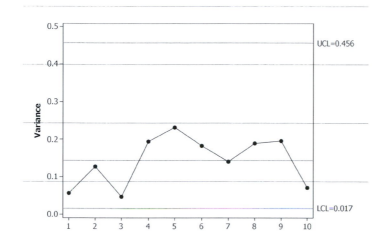

The process variation appears stable. No out-of-control signals.

e $\bar{\bar{y}} = 2.072$ $n = 9$ $m = 10$

$$LCL = \bar{\bar{y}} - 3\sqrt{\frac{\bar{s}^2}{n}} = 2.072 - 3(.126) = 1.69$$

$$UCL = \bar{\bar{y}} + 3\sqrt{\frac{\bar{s}^2}{n}} = 2.072 + 3(.126) = 2.45$$

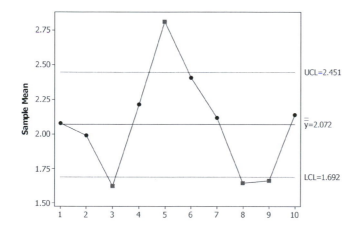

There are several out-of-control subgroups with a cyclical pattern.

f The s^2-chart assumes: 1) the process is in control during the base period, and 2) the data follow a normal distribution.

The $\overline{X}$-chart based on the sample variance assumes: 1) the s^2-chart is in control, 2) the subgroup means are in control during the base period, 3) the sample size is large enough to assume that the subgroup means follow a normal distribution by the Central Limit Theorem.

We should next look at a stem-and-leaf display of the data during the base period. The following display appears roughly normal.

```
Stem-and-leaf   N  = 90        Leaf Unit = 0.10

Stem             Leaves
1     0      7
2     0      9
3     1      1
7     1      2333
13    1      445555
28    1      666666666777777
41    1      8889999999999
      2      0000000000001111
33    2      222222223333
21    2      444555555
12    2      677
9     2      88999
4     3      01
2     3      3
1     3      4
```

g The process variation is stable in both a and d. However, there is much more variation when using cassette as the unit. Both charts for the mean show several out-of-control points with the same cyclical pattern.

5.25 a m = 30 observations in base period
$\mu_0 = 8$ microns $\overline{MR} = .2$

LCL = 8 − 2.66 (.2) = 7.468
UCL = 8 + 2.66 (.2) = 8.532

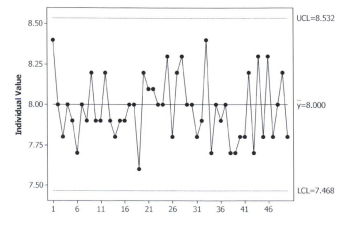

Process variation appears "in control"; no "out of control" signals.

b The X-chart assumes: 1) the process is in control during the base period, and 2) the data follow a normal distribution.

```
N = 30    Median = 8
Quartiles = 7.9, 8.1
    1    1    76 : 0
    2    1    77 : 0
    5    3    78 : 000
   12    7    79 : 0000000
         9    80 : 000000000
    9    2    81 : 00
    7    4    82 : 0000
    3    2    83 : 00
    1    1    84 : 0
```

Since this is a bell-shaped curve we can be comfortable with the normality assumption.

5.27 a m = 30 observations in base period

$\mu_0 = 0.2$ $\overline{MR} = .21$

$LCL = 0.2 - 2.66\,(.21) = -.359$

$UCL = 0.2 + 2.66\,(.21) = .759$

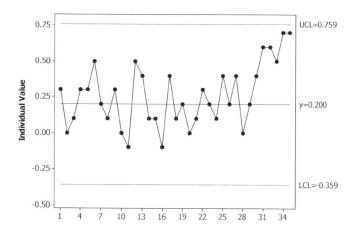

The instruments do not appear to agree. Also the process appears "out of control" for the last five observations.

b The X-chart assumes: 1) the process is in control during the base period, and 2) the data follow a normal distribution.

```
N = 30    Median = 0.2
Quartiles = 0.1, 0.3
Decimal point is 1 place to the left of the colon
    2    2   -1 : 00
    2    0   -0 :
    6    4    0 : 0000
   13    7    1 : 0000000
         5    2 : 00000
   12    5    3 : 00000
    7    5    4 : 00000
    2    2    5 : 00
```

This is single peaked with heavy tails, which makes us somewhat unsure about normality.

5.29 a $m = 30$

$\bar{\bar{y}} = 18.4$ $\overline{MR} = 3.34$

$LCL = 18.4 - 2.66(3.34) = 9.50$

$UCL = 18.4 + 2.66(3.34) = 27.3$

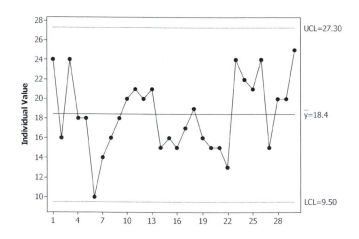

The process is stable, no out-of-control points.

b The X-chart assumes: 1) the process is in control during the base period, and 2) the data follow a normal distribution.

The stem-and-leaf display is inconclusive. A normal plot shows roughly a straight line. We should feel reasonably comfortable.

```
Stem-and-leaf of Iron  N  = 30      Leaf Unit = 1.0
Stem  Leaves
1     1  0
2     1  3
8     1  455555
13    1  66667
      1  8889
13    2  0000111
6     2  2
5     2  44445
```

5.31 a $m = 30$

$\bar{\bar{y}} = 4.37$ $\overline{MR} = 2.88$

$LCL = 4.37 - 2.66(2.88) = -3.29$

$UCL = 4.37 + 2.66(2.88) = 12.03$

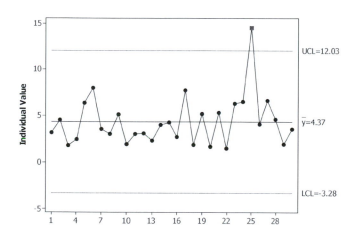

There is one out-of-control observation.

b The X-chart assumes: 1) the process is in control during the base period, and 2) the data follow a normal distribution.

The following display appears skewed right.

```
Stem-and-leaf of Concentration   N  = 30
Leaf Unit = 0.10

    6    1    468999
    9    2    347
   15    3    001156
   15    4    00256
   10    5    123
    7    6    3446
    3    7    7
    2    8    0
    1    9
    1   10
    1   11
    1   12
    1   13
    1   14    5
```

5.33 a n = 200 m = 30 subgroups in base period
$\bar{p} = .101$
$\text{LCL} = 200\,(.101) - 3\,\sqrt{200(.101)(.899)} = 7.416$
$\text{UCL} = 200\,(.101) + 3\,\sqrt{200(.101)(.899)} = 32.984$

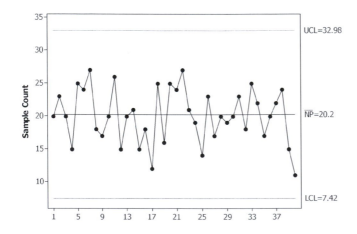

The hub caps from this particular supplier appear "in control". The incoming quality of hub caps appears stable.

5.35 a n = 300 m = 30 subgroups in base period

$\bar{p}$ = .0502

$$LCL = 300(.0502) - 3\sqrt{300(.0502)(.9497)} = 3.72$$

$$UCL = 300(.0502) + 3\sqrt{300(.0502)(.9497)} = 26.42$$

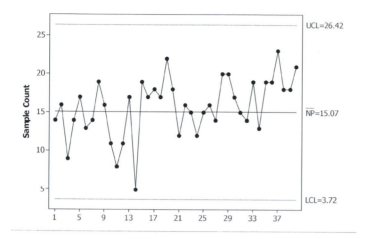

The writing instrument process appears "in control" with respect to proportion that should survive two years on a shelf.

5.37 a n = 200 m = 25 subgroups in base period

$\bar{p}$ = .0478

$$LCL = 200(.0478) - 3\sqrt{200(.0478)(.9522)} = .51$$

$$UCL = 200(.0478) + 3\sqrt{200(.0478)(.9522)} = 18.61$$

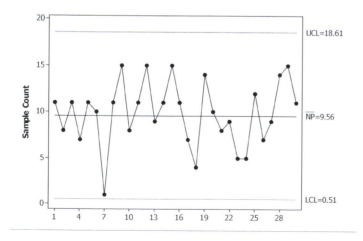

Production process appears "in control".

5.39 $n = 200$ $m = 20$

$\overline{p} = .0985$

$LCL = 200(.0985) - 3\sqrt{200(.0985)(.9015)} = 7.06$

$UCL = 200(.0985) + 3\sqrt{200(.0985)(.9015)} = 32.34$

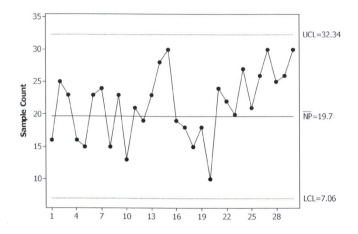

The process is stable, no out-of-control points.

5.41 a y_i number nonconforming is distributed as binomial (n_i, β)

$$\hat{p}_i = y_i/n_i \qquad E(\hat{p}_i) = p \qquad sd(\hat{p}_i) = \sqrt{p(1\text{-}p)/n_i}$$
$$LCL(\hat{p}_i) = p - 3\sqrt{p(1\text{-}p)/n_i} \qquad UCL(\hat{p}_i) = p + 3\sqrt{p(1\text{-}p)/n_i}$$

$\bar{p} = \Sigma y_i/m = \dfrac{\text{total number nonconforming}}{\text{total number inspected}}$. We will use $\bar{p}$ as an estimate for p.

$$LCL(\hat{p}_i) = \bar{p} - 3\sqrt{\bar{p}(1\text{-}\bar{p})/n_i} \qquad UCL(\hat{p}_i) = \bar{p} + 3\sqrt{\bar{p}(1\text{-}\bar{p})/n_i}$$

Assume the process is "in control" during base period, and use all 16 subgroups as a base period.
Then $\bar{p} = 278/3383 = .0674$

Subgroup	$\hat{p}_i$	LCL	UCL
1	.047	.02	.115
2	.039	.015	.120
3	.062	.019	.116
4	.0398	.014	.120
5	.078	.018	.117
6	.043	.003	.131
7	.055	.016	.118
8	.058	.007	.128
9	.063	.017	.118
10	.155	.015	.120
11	.090	.019	.115
12	.077	.017	.118
13	.075	.016	.119
14	.082	.019	.115
15	.051	.019	.116
16	.047	.006	.129

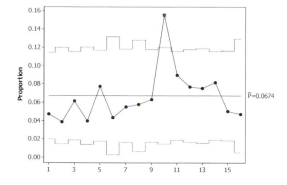

5.43 m = 20 subgroups in base period.

$\bar{c}$ = 5.7

LCL = $5.7 - 3\sqrt{5.7}$ = −1.46, so use 0

UCL = $5.7 + 3\sqrt{5.7}$ = 12.86

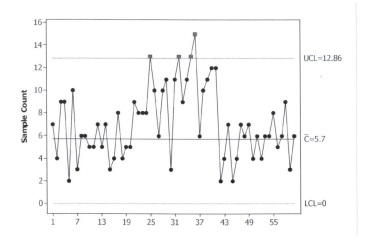

Process "in control" during base period. A number of "out of control" signals in subgroups 25-36. The process seems to become stable again after subgroup 36 process needs attention.

5.45 m = 30 subgroups in base period.

$\bar{c}$ = 11.367

LCL = $11.367 - 3\sqrt{11.367}$ = 1.25

UCL = $11.367 + 3\sqrt{11.367}$ = 21.48

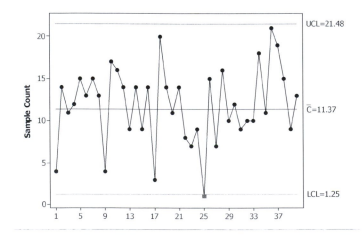

Subgroup 25 signals "out of control" and deserves attention.

5.47 m = 20 hours in base period

$\bar{c} = 9.8$

$LCL = 9.8 - 3\sqrt{9.8} = .41$

$UCL = 9.8 + 3\sqrt{9.8} = 19.2$

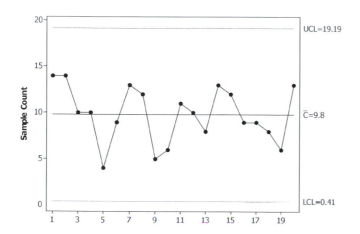

The production line appears stable over the past 20 hours of operation.

5.49 $m = 20$

$\bar{c} = 2.95$

$LCL = 2.95 - 3\sqrt{2.95} = 0$

$UCL = 2.95 + 3\sqrt{2.95} = 8.103$

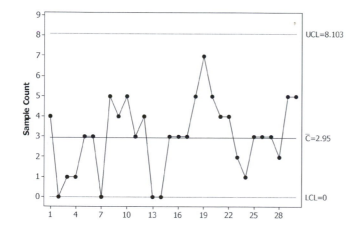

The process is stable, no out-of-control points.

5.51 $\mu = \mu_0$ $p(\mu_0) = .0027$ $\text{ARL}(\mu_0) = 1/.0027 = 370.37$

$\mu = \mu_0 + 0.5\sigma$ $p(\mu_0 + 0.5\sigma) = \Pr(Z < -4.5) + \Pr(Z > 1.5)$

$= .0668$ $\text{ARL}(\mu_0 + 0.5\sigma) = 1/.0668 = 14.97$

$\mu = \mu_0 + \sigma$ $p(\mu_0 + \sigma) = \Pr(Z < -6) + \Pr(Z > 0)$

$= .5$ $\text{ARL}(\mu_0 + \sigma) = 1/.5 = 2.00$

$\mu = \mu_0 + 1.5\sigma$ $p(\mu_0 + 1.5\sigma) = \Pr(Z < -7.5) + \Pr(Z > 1.5)$

$= .9332$ $\text{ARL}(\mu_0 + 1.5\sigma) = 1/.9332 = 1.07$

$\mu = \mu_0 + 2\sigma$ $p(\mu_0 + 2\sigma) = \Pr(Z < -9) + \Pr(Z > -3)$

$= .9987$ $\text{ARL}(\mu_0 + 2\sigma) = 1/.9987 = 1.00$

$\mu = \mu_0 + 2.5\sigma$ $p(\mu_0 + 2.5\sigma) = \Pr(Z < -10.5) + \Pr(Z > -4.5)$

$= 1$ $\text{ARL}(\mu_0 + 2.5\sigma) = 1/1 = 1.00$

5.53

$n = 5$

μ	$p(\mu)$	$\text{ARL}(\mu)$
μ_0	.0027	370.37
$\mu_0 + 0.5\sigma$	.0301	33.22
$\mu_0 + \sigma$	.2236	4.47
$\mu_0 + 1.5\sigma$	.6368	1.57
$\mu_0 + 2\sigma$	.9292	1.08
$\mu_0 + 2.5\sigma$	.9952	1.00

5.55 $\mu_0 = 100$ $\sigma = 8$ $\sigma_\theta = 8/\sqrt{4}$
LCL = 88   UCL = 112
warning limits:
$$100 + 2(8/\sqrt{4}) = 108$$
$$100 + (8/\sqrt{4}) = 104$$
$$100 - (8/\sqrt{4}) = 96$$
$$100 - 2(8/\sqrt{4}) = 92$$

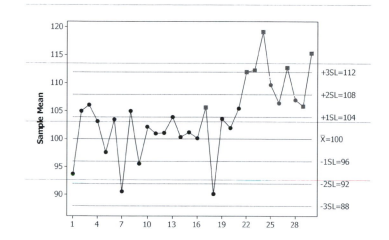

rule 1   no signals
rule 2   signal with subgroup 29
subgroups 25-29 in zone 2 or 3
rule 3   signal with subgroup 17 subgroups
          10-17 in zones 2, 3, 4

Runs rule 3 helps for chart to signal earlier than regular $\overline{X}$ -chart.

5.57 $\bar{R} = .44$ $n = 4$

$LCL = D_3\bar{R} = 0(.44) = 0 = (1 + 3d_3/d_2)\bar{R} = (1 - 3 (.880/2.059))(.44)$

$UCL = D_4\bar{R} = 2.282(.44) = 1.00 = (1 - 3d_3/d_2)\bar{R} = (1 + 3 (.880/2.059))(.44)$

warning limits:

$(1 + 2d_3/d_2)\bar{R} = (1 + 2(.427))(.44) = .816$

$(1 + d_3/d_2)\bar{R} = (1 + (.427))(.44) = .628$

$(1 - d_3/d_2)\bar{R} = (1 - .427)(.44) = .252$

$(1 - 2d_3/d_2)\bar{R} = (1 - 2(.427))(.44) = .064$

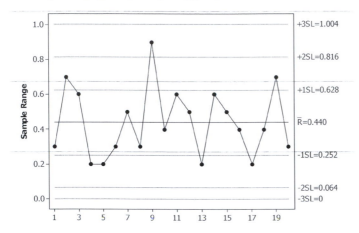

rule 1   no signals

rule 2   no signals

rule 3   no signals   process variation "in control"

$\bar{\bar{y}} = 19.388$ $n = 4$

$LCL = \bar{\bar{y}} - A_2\bar{R} = 19.388 - .729(.44) = 19.07 = \bar{\bar{y}} - 3\bar{R} / d_2\sqrt{4} = 19.388 - 3(.44)/2.059(2)$

$UCL = \bar{\bar{y}} - A_2\bar{R} = 19.388 + .729(.44) = 19.71 = \bar{\bar{y}} + 3\bar{R} / d_2\sqrt{4} = 19.388 + 3(.44)/2.059(2)$

warning limits:

$\bar{\bar{y}} + 2\bar{R} / d_2\sqrt{4} = 19.388 + \frac{2(.44)}{2.059(2)} = 19.60$

$\bar{\bar{y}} + \bar{R} / d_2\sqrt{4} = 19.388 + \frac{(.44)}{2.059(2)} = 19.495$

$\bar{\bar{y}} - \bar{R} / d_2\sqrt{4} = 19.388 - \frac{(.44)}{2.059(2)} = 19.28$

$\bar{\bar{y}} - 2\bar{R} / d_2\sqrt{4} = 19.388 - \frac{2(.44)}{2.059(2)} = 19.17$

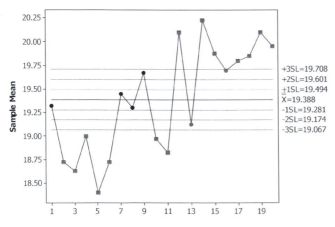

Process clearly "out of control" -- runs rules are not much help.

5.59 $\overline{p} = .00979$

$\qquad$ LCL = 2.89 $\qquad\qquad$ UCL = 25.31

$\qquad$ warning limits $n\overline{p} \pm 2\sqrt{n\overline{p}(1-\overline{p})}$ and $n\overline{p} \pm \sqrt{n\overline{p}(1-\overline{p})}$

$\qquad\qquad 1440(.00979) + 2\sqrt{1440(.00979)(.99021)} = 21.57$

$\qquad\qquad 1440(.00979) + \sqrt{1440(.00979)(.99021)} = 17.84$

$\qquad\qquad 1440(.00979) - \sqrt{1440(.00979)(.99021)} = 10.36$

$\qquad\qquad 1440(.00979) - 2\sqrt{1440(.00979)(.99021)} = 6.63$

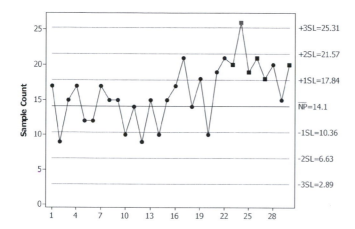

rule 1   no signals

rule 2   signal for subgroup 23 subgroups 19-23

$\qquad\qquad$ signal for subgroup 25 subgroups 21-25

$\qquad\qquad$ signal for subgroup 26 subgroups 22-26

$\qquad\qquad$ signal for subgroup 27 subgroups 23-27

$\qquad\qquad$ signal for subgroup 30 subgroups 26-30

Rules detect shift quicker than np chart.

5.61 $\bar{c} = 5.7$

    $LCL = 0$ $UCL = 12.86$

      warning limits $\bar{c} \pm 2\sqrt{\bar{c}}$ and $\bar{c} \pm \sqrt{\bar{c}}$

        $5.7 + 2\sqrt{5.7} = 10.47$

        $5.7 + \sqrt{5.7} = 8.09$

        $5.7 - \sqrt{5.7} = 3.31$

        $5.7 - 2\sqrt{5.7} = .93$

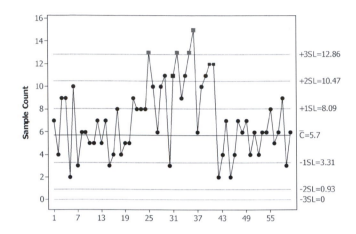

rule 1   First signal for subgroup 31, by subgroups 29-31 in zone 2

        Also signals for subgroups 40, 41

rule 2   Signal for subgroup 41

Not much gained from runs rules since the process is very unstable.

5.63 $m = 20$ $\bar{c} = 7.1$

$LCL = 7.1 - 3\sqrt{7.1} = 0$ $UCL = 7.1 + 3\sqrt{7.1} = 15.09$

warning limits:

$7.1 + 2\sqrt{7.1} = 12.43$

$7.1 + \sqrt{7.1} = 9.76$

$7.1 - \sqrt{7.1} = 4.44$

$7.1 - 2\sqrt{7.1} = 1.77$

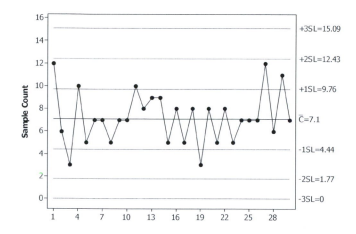

rule 1   no signals
rule 2   no signals

5.65 a $\bar{y} = .7336$ $\overline{MR} = .048$ $\hat{\sigma} = .0425$

$Z_i = (y_i - .7336)/.0425$

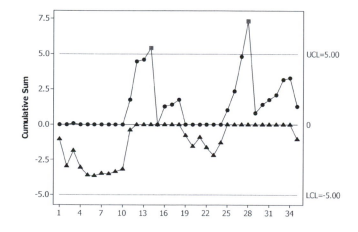

The CUSUM chart for $\mu_1 > \mu_0$ indicates the process location is increasing.

b $m = 25$ $\bar{\bar{y}} = .7336$ $\overline{MR} = .048$ $Z_i = .1X_i + .9Z_{i-1}$

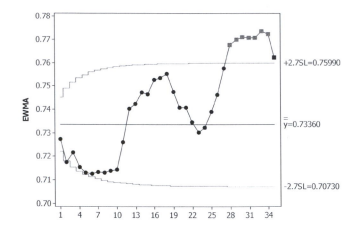

c   All three charts have out-of-control signals, but the EWMA shows an out-of-control pattern at the end of the time period.

5.67 **a** $\bar{y} = 16.52$ $\overline{MR} = .126$ $\hat{\sigma} = .112$
$Z_i = (y_i - 16.52)/.112$

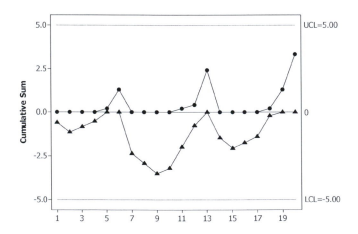

The chart does not signal, but has a cyclical trend.

b $m = 20$ $\bar{\bar{y}} = 16.52$ $\overline{MR} = .126$ $Z_i = .1X_i + .9Z_{i-1}$

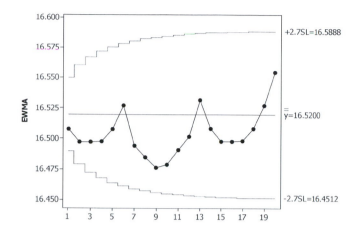

c All three charts show a stable process.

5.69 a $\mu_0 = 4.374$ $\overline{MR} = 2.88$ $\hat{\sigma} = 2.553$
$Z_i = (y_i - 4.374)/2.553$

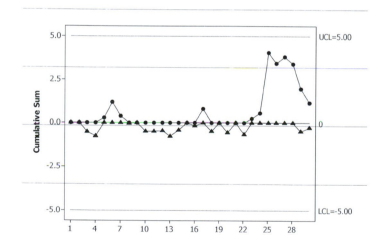

The chart does not signal.

b $m = 30$ $\bar{\bar{y}} = 4.374$ $\overline{MR} = 2.88$ $Z_i = .1X_i + .9Z_{i-1}$

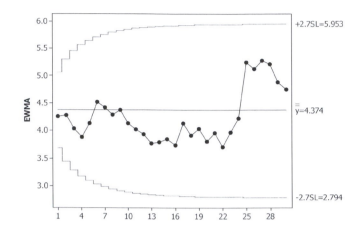

c   The CUSUM and EWMA charts show a stable process, but the X-chart had one out-of-control signal.

5.71 a $\bar{s} = 2.82443$ $\bar{X} = 34.0067$ LSL = 25   USL = 45

$C_p = (45 - 25)/6 \cdot 2.82443 = 1.18$

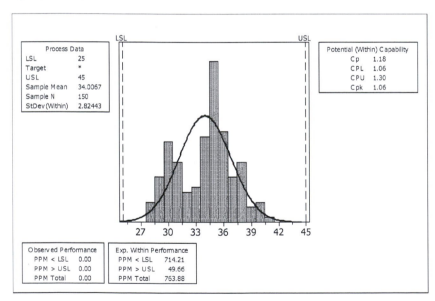

b   CPU = $(45 - 34.0067)/3 \cdot 2.82443 = 1.30$,   CPL = $(34.0067 - 25)/3 \cdot 2.82443 = 1.06$

$C_{pk} = \min(\text{CPL, CPU}) = 1.06$

c   Because the mean of the data is off from the center of the specification limits, C_p and C_{pk} are different.

d $PPM_U = Pr(X > 45) = Pr\left(\dfrac{X - \bar{X}}{\bar{s}} > \dfrac{45 - 34.0067}{2.82443}\right)$

$\qquad = Pr(Z > 3.89) = .0000501$

$PPM_L = Pr(X < 25) = Pr\left(\dfrac{X - \bar{X}}{\bar{s}} < \dfrac{25 - 34.0067}{2.82443}\right)$

$\qquad = Pr(Z < -3.19) = .0007114$

$1,000,000 \cdot PPM_U + 1,000,000 \cdot PPM_L = 761.5$ (off from the graph above due to rounding) indicating the process is not too good.

5.73 a $\bar{s} = 1.61385$ $\qquad \bar{X} = 261.191$ $\qquad LSL = 250$ $\qquad USL = 270$

$\qquad C_p = (270 - 250)/6 \cdot 1.61385 = 2.07$

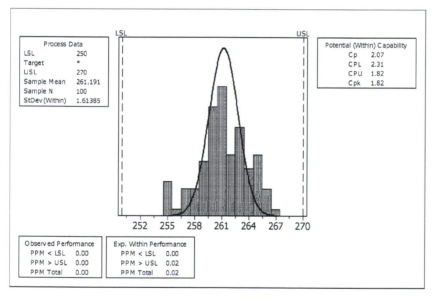

b $CPU = (270 - 261.191)/3 \cdot 1.61385 = 1.82$, $CPL = (261.191 - 250)/3 \cdot 1.61385 = 2.31$

$\qquad C_{pk} = \min(CPL, CPU) = 1.82$

c Because the mean of the data is off from the center of the specification limits, C_p and C_{pk} are different.

d $PPM_U = Pr(X > 270) = Pr\left(\dfrac{X - \bar{X}}{\bar{s}} > \dfrac{270 - 261.191}{1.61385}\right)$

$\qquad = Pr(Z > 5.46) \approx 0$

$PPM_L = Pr(X < 250) = Pr\left(\dfrac{X - \bar{X}}{\bar{s}} < \dfrac{250 - 261.191}{1.61385}\right)$

$\qquad = Pr(Z < -6.93) \approx 0$

$1,000,000 \cdot PPM_U + 1,000,000 \cdot PPM_L \approx 0$ indicating the process is very good.

5.75 a $\bar{s} = 2.85597$ $\bar{X} = 19.152$ USL $= 30$

$C_{pk} = (30 - 19.152)/3 \cdot 2.85597 = 1.27$

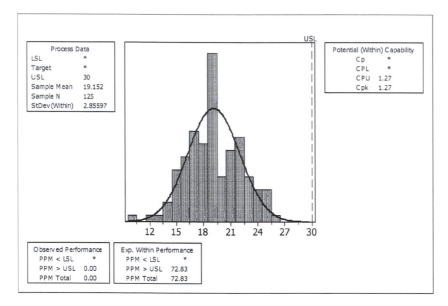

b $\text{PPM}_U = \Pr(X > 30) = \Pr\left(\dfrac{X - \bar{X}}{\bar{s}} > \dfrac{30 - 19.152}{2.85597} \right)$

$= \Pr(Z > 3.80) = .0000723$

$1{,}000{,}000 \cdot \text{PPM}_L = 72.3$ (off from the graph above due to rounding), indicating the process is reasonably good.

5.77

Source	StdDev (SD)	Study Var (6 * SD)	%Study Var (%SV)
Total Gage R&R	0.023477	0.14086	5.74
Repeatability	0.022464	0.13478	5.49
Reproducibility	0.006824	0.04094	1.67
Part-To-Part	0.408482	2.45089	99.84
Total Variation	0.409156	2.45494	100.00

a 0.022464

b 0.006824

c 0.408482

d 5.74%, which is acceptable.

5.79

Source	StdDev (SD)	Study Var (6 * SD)	%Study Var (%SV)
Total Gage R&R	0.382271	2.29362	85.00
Repeatability	0.339655	2.03793	75.53
Reproducibility	0.175400	1.05240	39.00
Part-To-Part	0.236895	1.42137	52.68
Total Variation	0.449722	2.69833	100.00

a 0.339655

b 0.1754

c 0.236895

d 85%, which is unacceptable. The measurement system variation is mostly repeatability.

5.81

Source	StdDev (SD)	Study Var (6 * SD)	%Study Var (%SV)
Total Gage R&R	0.2628	1.5766	2.30
Repeatability	0.1378	0.8269	1.20
Reproducibility	0.2237	1.3423	1.96
Part-To-Part	11.4396	68.6373	99.97
Total Variation	11.4426	68.6554	100.00

a 0.1378

b 0.2237

c 11.4396

d 2.3%, which is acceptable.

CHAPTER 6

LINEAR REGRESSION ANALYSIS

6.1 Dependent Variable: DAYS = number of days the ozone levels exceeded 0.20 ppm
Independent Variable: INDEX = seasonal meteorological index

Analysis of Variance

Source	DF	Sum of Squares	Mean Square	F Value	Prob>F
Model	1	1492.63844	1492.63844	2.636	0.1267
Error	14	7926.79906	566.19993		
C Total	15	9419.43750			

Root MSE	23.79496	R-square	0.1585	
Dep Mean	72.31250	Adj R-sq	0.0984	
C.V.	32.90573			

Parameter Estimates

Variable	DF	Parameter Estimate	Standard Error	T for H0: Parameter=0	Prob > |T|
INTERCEP	1	-192.983835	163.50328346	-1.180	0.2575
INDEX	1	15.296365	9.42097477	1.624	0.1267

PREDICTION EQUATION: days = −192.983835 + 15.296365(index)

$R^2 = 0.1585$ and Adj $R^2 = 0.0984$ are both very low values. This model does not fit the data very well. The F-test, F = 2.636, is not larger than $F(1,14,.05) = 4.60$ and p-value = .1267. The T-test of index, T = 1.624, is not larger than $T(14, .025) = 2.145$ and p-value = .1267. We cannot conclude that index and days are related.

Obs	Dep Var DAYS	Predict Value	Std Err Predict	Lower95% Mean	Upper95% Mean	Lower95% Predict	Upper95% Predict
1	91.0000	62.4655	8.495	44.2450	80.6859	8.2754	116.7
2	105.0	68.5840	6.377	54.9076	82.2604	15.7482	121.4
3	106.0	85.4100	10.023	63.9129	106.9	30.0322	140.8
4	108.0	83.8804	9.282	63.9734	103.8	29.1002	138.7
5	88.0000	70.1136	6.101	57.0284	83.1989	17.4277	122.8
6	91.0000	85.4100	10.023	63.9129	106.9	30.0322	140.8
7	58.0000	51.7580	13.987	21.7579	81.7581	-7.4416	111.0
8	82.0000	70.1136	6.101	57.0284	83.1989	17.4277	122.8
9	81.0000	82.3507	8.580	63.9491	100.8	28.0995	136.6
10	65.0000	70.1136	6.101	57.0284	83.1989	17.4277	122.8
11	61.0000	65.5247	7.271	49.9304	81.1191	12.1603	118.9
12	48.0000	68.5840	6.377	54.9076	82.2604	15.7482	121.4
13	61.0000	85.4100	10.023	63.9129	106.9	30.0322	140.8
14	43.0000	71.6433	5.963	58.8539	84.4326	19.0301	124.3
15	33.0000	74.7026	6.128	61.5590	87.8462	22.0021	127.4
16	36.0000	60.9358	9.191	41.2220	80.6496	6.2255	115.6

123

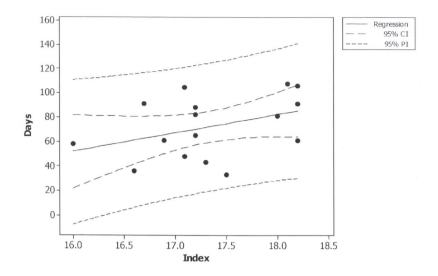

6.3   Dependent Variable: VISCOS = intrinsic viscosity
Independent Variable: RATIO = molar ratio of sebacic acid

Analysis of Variance

Source	DF	Sum of Squares	Mean Square	F Value	Prob>F
Model	1	0.03691	0.03691	1.640	0.2475
Error	6	0.13498	0.02250		
C Total	7	0.17189			

Root MSE	0.14999	R-square	0.2147	
Dep Mean	0.47875	Adj R-sq	0.0838	
C.V.	31.32952			

Parameter Estimates

Variable	DF	Parameter Estimate	Standard Error	T for H0: Parameter=0	Prob > |T|
INTERCEP	1	0.671429	0.15950882	4.209	0.0056
RATIO	1	-0.296429	0.23143972	-1.281	0.2475

PREDICTION EQUATION: viscosity = 0.671429 − 0.296429(ratio)

$R^2 = 0.2147$ and Adj $R^2 = 0.0838$ are both very low values. This model does not fit the data very well. The F-test, F = 1.640, is not larger than F(1, 6, .05) = 5.99. The T-test of ratio, T = −1.281, is not smaller than −T(6, .025) = −2.447. The p-value for both tests = .2475 is large. We cannot conclude that the viscosity of copolyesters and the molar ratio of sebacic acid are related.

Obs	Dep Var VISCOS	Predict Value	Std Err Predict	Lower95% Mean	Upper95% Mean	Lower95% Predict	Upper95% Predict
1	0.4500	0.3750	0.097	0.1381	0.6119	-0.0618	0.8118
2	0.2000	0.4046	0.078	0.2126	0.5967	-0.00958	0.8189
3	0.3400	0.4343	0.063	0.2792	0.5894	0.0358	0.8327
4	0.5800	0.4639	0.054	0.3311	0.5967	0.0736	0.8542
5	0.7000	0.4936	0.054	0.3608	0.6264	0.1033	0.8839
6	0.5700	0.5232	0.063	0.3681	0.6783	0.1248	0.9217
7	0.5500	0.5529	0.078	0.3608	0.7449	0.1386	0.9671
8	0.4400	0.5825	0.097	0.3456	0.8194	0.1457	1.0193

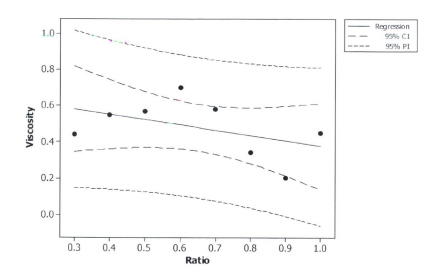

6.5   Dependent Variable: VIS = viscosity of toluene-tetralin blends
Independent Variable: TEMP = temperature

Analysis of Variance

Source	DF	Sum of Squares	Mean Square	F Value	Prob>F
Model	1	0.32529	0.32529	144.578	0.0001
Error	6	0.01350	0.00225		
C Total	7	0.33879			

Root MSE	0.04743	R-square	0.9602	
Dep Mean	0.75528	Adj R-sq	0.9535	
C.V.	6.28031			

Parameter Estimates

Variable	DF	Parameter Estimate	Standard Error	T for H0: Parameter=0	Prob > |T|
INTERCEP	1	1.281511	0.04686833	27.343	0.0001
TEMP	1	-0.008758	0.00072836	-12.024	0.0001

PREDICTION EQUATION: viscosity = 1.281511 − 0.008758(temperature)

$R^2 = 0.9602$ and Adj $R^2 = 0.9535$ are both larger than 0.9. This model fits the data very well. The F-test, F = 144.578, is larger than F(1,6, .05) = 5.99. The T-test of temp, T = −12.024, is smaller than −T(6, .025) = −2.447. The p-value for both tests is .0001 very small. We can conclude that the temperature does impact the viscosity of toluene-tetralin blends.

Obs	Dep Var VIS	Predict Value	Std Err Predict	Lower95% Mean	Upper95% Mean	Lower95% Predict	Upper95% Predict
1	1.1330	1.0634	0.031	0.9885	1.1384	0.9253	1.2016
2	0.9772	0.9750	0.025	0.9143	1.0357	0.8440	1.1060
3	0.8532	0.8883	0.020	0.8391	0.9374	0.7622	1.0143
4	0.7550	0.7990	0.017	0.7570	0.8409	0.6755	0.9224
5	0.6723	0.7105	0.017	0.6685	0.7525	0.5871	0.8339
6	0.6021	0.6229	0.020	0.5738	0.6720	0.4969	0.7489
7	0.5420	0.5353	0.025	0.4746	0.5961	0.4044	0.6663
8	0.5074	0.4478	0.031	0.3729	0.5226	0.3097	0.5859

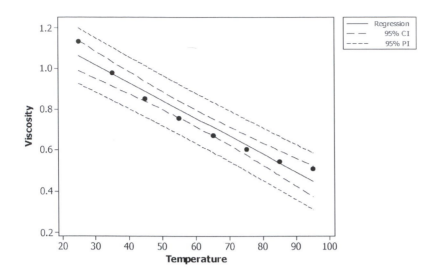

6.7   Dependent Variable: WEAR = wear on the cylinder blocks
Independent Variable: DISCHARGE = discharge of oil

Analysis of Variance

Source	DF	Sum of Squares	Mean Square	F Value	Prob>F
Model	1	0.04629	0.04629	13553.482	0.0001
Error	6	0.0000204936	3.415608E-6		
C Total	7	0.04631			

Root MSE	0.00185	R-square	0.9996	
Dep Mean	0.12238	Adj R-sq	0.9995	
C.V.	1.51022			

Parameter Estimates

Variable	DF	Parameter Estimate	Standard Error	T for H0: Parameter=0	Prob > |T|
INTERCEP	1	0.300613	0.00166461	180.591	0.0001
DISCHARG	1	-0.025926	0.00022269	-116.419	0.0001

PREDICTION EQUATION: Wear = 0.300613 − 0.025926(discharge)

$R^2 = 0.9996$ and Adj $R^2 = 0.9995$ are both close to one. This model fits the data extremely well. The F-test value, F = 13553.482, is larger than F(1,6, .05) = 5.99. The T-test of discharge, T = −116.419, is smaller than −T(6, .025) = −2.447. Strong evidence of relationship since p-value is 0.0001. We can conclude that there is strong relationship discharge of oil on the wear on the cylinder between the blocks for a specific type of hydraulic pump.

Obs	Dep Var WEAR	Predict Value	Std Err Predict	Lower95% Mean	Upper95% Mean	Lower95% Predict	Upper95% Predict
1	0.0160	0.0154	0.001	0.0127	0.0182	0.0101	0.0207
2	0.0400	0.0414	0.001	0.0390	0.0437	0.0363	0.0464
3	0.0670	0.0673	0.001	0.0653	0.0693	0.0623	0.0722
4	0.0960	0.0932	0.001	0.0915	0.0949	0.0884	0.0980
5	0.1450	0.1451	0.001	0.1434	0.1467	0.1402	0.1499
6	0.1680	0.1710	0.001	0.1691	0.1729	0.1661	0.1759
7	0.1970	0.1969	0.001	0.1947	0.1991	0.1919	0.2020
8	0.2500	0.2488	0.001	0.2457	0.2519	0.2433	0.2542

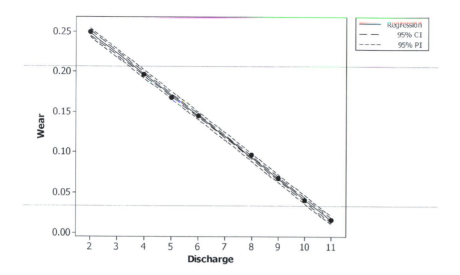

6.9 Dependent Variable: Per. Noncontam = percent noncontaminated
Independent Variable: Aflatoxin

The regression equation is
Perc. Noncontam. = 100 − 0.00290 Aflatoxin

```
Predictor          Coef      SE Coef         T       P
Constant        100.002        0.011   9184.91   0.000
Aflatoxin    -0.0029035    0.0002335    -12.43   0.000
```

S = 0.0393282   R-Sq = 82.9%   R-Sq(adj) = 82.3%

Analysis of Variance
```
Source           DF       SS        MS        F       P
Regression        1  0.23915   0.23915   154.62   0.000
Residual Error   32  0.04949   0.00155
Total            33  0.28864
```

R^2 = .829 and Adj R^2 = .823 are both high values. This model fits the data pretty well. The F-test value, F = 154.62 with a p-value of 0.000. We can conclude that the average level of aflatoxin and the percentage of noncontaminated peanuts are related.

```
Dep Var   Fit       SE Fit       95% CI               95% PI
99.971    99.9934   0.0103    (99.9723, 100.0145)   (99.9106, 100.0762)
99.979    99.9885   0.0100    (99.9680, 100.0089)   (99.9058, 100.0711)
99.982    99.9780   0.0094    (99.9588,  99.9972)   (99.8956, 100.0604)
99.971    99.9751   0.0093    (99.9562,  99.9940)   (99.8928, 100.0574)
99.957    99.9734   0.0092    (99.9546,  99.9921)   (99.8911, 100.0556)
99.961    99.9702   0.0090    (99.9518,  99.9885)   (99.8880, 100.0523)
99.956    99.9664   0.0088    (99.9484,  99.9843)   (99.8843, 100.0485)
99.972    99.9658   0.0088    (99.9479,  99.9837)   (99.8837, 100.0479)
99.889    99.9655   0.0088    (99.9477,  99.9834)   (99.8834, 100.0476)
99.961    99.9559   0.0083    (99.9390,  99.9728)   (99.8741, 100.0378)
99.982    99.9536   0.0082    (99.9369,  99.9703)   (99.8718, 100.0354)
99.975    99.9475   0.0079    (99.9314,  99.9637)   (99.8658, 100.0292)
99.942    99.9475   0.0079    (99.9314,  99.9637)   (99.8658, 100.0292)
99.932    99.9472   0.0079    (99.9311,  99.9633)   (99.8655, 100.0289)
99.908    99.9391   0.0076    (99.9236,  99.9546)   (99.8575, 100.0207)
99.970    99.9385   0.0076    (99.9231,  99.9539)   (99.8569, 100.0201)
99.985    99.9359   0.0075    (99.9207,  99.9511)   (99.8544, 100.0174)
99.933    99.9318   0.0073    (99.9169,  99.9468)   (99.8503, 100.0133)
99.858    99.9272   0.0072    (99.9125,  99.9419)   (99.8457, 100.0086)
99.987    99.9133   0.0069    (99.8992,  99.9273)   (99.8319,  99.9946)
99.958    99.8970   0.0067    (99.8833,  99.9107)   (99.8157,  99.9783)
99.909    99.8865   0.0068    (99.8727,  99.9004)   (99.8052,  99.9678)
99.859    99.8735   0.0070    (99.8593,  99.8877)   (99.7921,  99.9548)
99.863    99.8662   0.0072    (99.8516,  99.8808)   (99.7848,  99.9476)
99.811    99.8662   0.0072    (99.8516,  99.8808)   (99.7848,  99.9476)
99.877    99.8334   0.0084    (99.8163,  99.8505)   (99.7515,  99.9153)
99.798    99.8212   0.0090    (99.8028,  99.8396)   (99.7390,  99.9034)
99.855    99.7971   0.0104    (99.7759,  99.8183)   (99.7142,  99.8800)
99.788    99.7957   0.0105    (99.7743,  99.8171)   (99.7127,  99.8786)
99.821    99.7951   0.0105    (99.7736,  99.8166)   (99.7121,  99.8780)
99.830    99.7605   0.0128    (99.7345,  99.7866)   (99.6763,  99.8448)
99.718    99.7594   0.0129    (99.7331,  99.7856)   (99.6751,  99.8437)
99.642    99.7132   0.0162    (99.6803,  99.7461)   (99.6266,  99.7998)
99.658    99.6792   0.0187    (99.6412,  99.7173)   (99.5905,  99.7679)
```

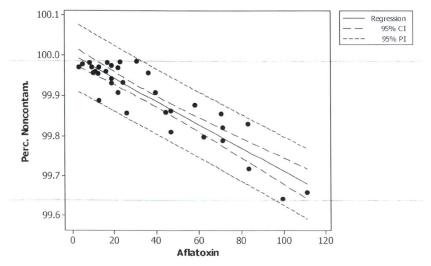

6.11 Dependent Variable: Per. Absorbed% = percent absorbed
Independent Variable: Amount

The regression equation is
Absorbed % = −1822 + 435 Amount

```
Predictor      Coef   SE Coef      T      P
Constant    -1821.8     365.7  -4.98  0.000
Amount       434.73     60.25   7.22  0.000
```

S = 204.898   R-Sq = 70.3%   R-Sq(adj) = 68.9%

Analysis of Variance

```
Source            DF       SS       MS      F      P
Regression         1  2185973  2185973  52.07  0.000
Residual Error    22   923627    41983
Total             23  3109600
```

$R^2 = .703$ and Adj $R^2 = .689$ are both moderate values. This model fit the data only moderately well. The F-test value, F = 52.07 with a p-value of 0.000. We can conclude that the amount of liquid and the absorbed percentage are related.

```
Dep Var    Fit    SE Fit      95% CI            95% PI
310      143.2    100.2  ( -64.6,   350.9)  (-329.8,   616.2)
330      430.1     66.2  ( 292.9,   567.3)  ( -16.4,   876.6)
370      682.3     44.9  ( 589.2,   775.4)  ( 247.2,  1117.3)
400      395.3     70.0  ( 250.2,   540.4)  ( -53.7,   844.4)
450      825.7     42.0  ( 738.7,   912.8)  ( 392.0,  1259.5)
490      591.0     50.9  ( 485.5,   696.5)  ( 153.1,  1028.8)
520      551.8     54.1  ( 439.5,   664.1)  ( 112.3,   991.4)
560      699.6     44.1  ( 608.2,   791.1)  ( 265.0,  1134.3)
580      569.2     52.7  ( 460.0,   678.4)  ( 130.5,  1008.0)
650      460.5     62.9  ( 330.0,   591.1)  (  16.0,   905.1)
650      838.8     42.2  ( 751.3,   926.2)  ( 404.9,  1272.6)
650      751.8     42.4  ( 664.0,   839.7)  ( 317.9,  1185.7)
760     1060.5     55.2  ( 945.9,  1175.1)  ( 620.4,  1500.6)
800      560.5     53.4  ( 449.8,   671.3)  ( 121.4,   999.7)
810      882.2     43.3  ( 792.3,   972.1)  ( 447.9,  1316.6)
910      734.4     42.8  ( 645.7,   823.2)  ( 300.3,  1168.5)
1020    1217.0     71.3  (1069.0,  1364.9)  ( 767.0,  1666.9)
1020     917.0     44.9  ( 824.0,  1010.0)  ( 482.0,  1352.0)
1160    1160.5     65.2  (1025.3,  1295.6)  ( 714.6,  1606.4)
1200    1104.0     59.4  ( 980.8,  1227.1)  ( 661.5,  1546.4)
1230     951.8     46.8  ( 854.7,  1048.9)  ( 515.9,  1387.7)
1380    1295.2     80.4  (1128.6,  1461.9)  ( 838.8,  1751.7)
1460    1321.3     83.5  (1148.2,  1494.4)  ( 862.5,  1780.2)
1490    1056.1     54.9  ( 942.4,  1169.9)  ( 616.2,  1496.0)
```

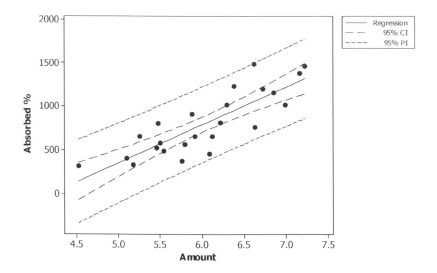

6.13   Dependent Variable: Attenuation
Independent Variable: Tensile strength

The regression equation is
Attenuation = 3.62 −0.0147 Tensile Strength

Predictor	Coef	SE Coef	T	P
Constant	3.62091	0.09949	36.39	0.000
Tensile Strength	-0.014711	0.001436	-10.24	0.000

S = 0.147890   R-Sq = 89.7%   R-Sq(adj) = 88.9%

Analysis of Variance

Source	DF	SS	MS	F	P
Regression	1	2.2947	2.2947	104.92	0.000
Residual Error	12	0.2625	0.0219		
Total	13	2.5571			

$R^2 = .897$ and Adj $R^2 = .889$ are both high values. This model fit the data well. The F-test value, F = 104.92 with a p-value of 0.000. We can conclude that the percentage of ultimate tensile strength and the decrease in amplitude of the stress wave are related.

Dep Var	Fit	SE Fit	95% CI	95% PI
3.3	3.4444	0.0840	(3.2615, 3.6273)	(3.0739, 3.8149)
3.2	3.1796	0.0623	(3.0437, 3.3154)	(2.8299, 3.5293)
3.4	3.0913	0.0559	(2.9694, 3.2132)	(2.7468, 3.4358)
3.0	3.0325	0.0520	(2.9191, 3.1459)	(2.6909, 3.3741)
2.8	2.9589	0.0477	(2.8550, 3.0628)	(2.6204, 3.2975)
2.9	2.7824	0.0406	(2.6938, 2.8709)	(2.4482, 3.1166)
2.7	2.7088	0.0396	(2.6226, 2.7951)	(2.3753, 3.0424)
2.6	2.6353	0.0398	(2.5485, 2.7221)	(2.3016, 2.9690)
2.5	2.5764	0.0409	(2.4872, 2.6656)	(2.2421, 2.9108)
2.6	2.4735	0.0446	(2.3762, 2.5707)	(2.1369, 2.8100)
2.2	2.2528	0.0579	(2.1267, 2.3789)	(1.9068, 2.5988)
2.0	2.2381	0.0589	(2.1097, 2.3665)	(1.8912, 2.5849)
2.3	2.1498	0.0656	(2.0070, 2.2927)	(1.7973, 2.5023)
2.1	2.0763	0.0714	(1.9206, 2.2319)	(1.7184, 2.4341)

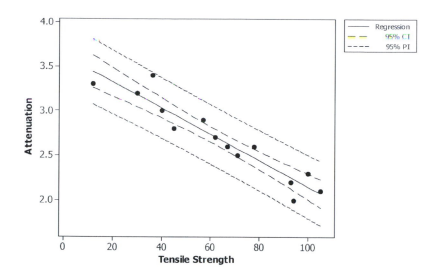

6.15 Dependent Variable: DIFF = temperature difference on the die surface
Independent Variables: TEMP = furnace temperature
TIME = die close time

Analysis of Variance

Source	DF	Sum of Squares	Mean Square	F Value	Prob>F
Model	2	715.50000	357.75000	319.314	0.0001
Error	6	6.72222	1.12037		
C Total	8	722.22222			

Root MSE	1.05848	R-square	0.9907	
Dep Mean	94.44444	Adj R-sq	0.9876	
C.V.	1.12074			

Parameter Estimates

Variable	DF	Parameter Estimate	Standard Error	T for H0: Parameter=0	Prob > |T|
INTERCEP	1	-199.555556	11.64055720	-17.143	0.0001
TEMP	1	0.210000	0.00864242	24.299	0.0001
TIME	1	3.000000	0.43212081	6.943	0.0004

PREDICTION EQUATION: Diff = -199.555556 + 0.21(temp) + 3(time)

R^2 = .9907 and Adj R^2 = .9876 are both very close to one. This model fits the data very well. The F-test value, F = 319.314 is larger than F(2,6, .05) = 5.14. The T-test of temp, T = 24.299 is also a lot larger than T(6, .025) = 2.447. The T-test of time, T = 6.943 is also larger than T(6, .025) = 2.447. The overall F-test p-value = 0.0001 and the p-values for both predictors are less than 0.05. We can conclude that the furnace temperature and the die close time have positive significant effect on the temperature difference on the die surface in a die casting process. In order to minimize the difference, we should use lower temperature and time.

6.17 Dependent Variable: Y
Independent Variables: X1, X2, X3

Analysis of Variance

Source	DF	Sum of Squares	Mean Square	F Value	Prob>F
Model	3	930.50000	310.16667	69.920	0.0001
Error	24	106.46429	4.43601		
C Total	27	1036.96429			

Root MSE	2.10618	R-square	0.8973
Dep Mean	80.53571	Adj R-sq	0.8845
C.V.	2.61522		

Parameter Estimates

Variable	DF	Parameter Estimate	Standard Error	T for H0: Parameter=0	Prob > |T|
INTERCEP	1	80.535714	0.39803140	202.335	0.0001
X1	1	-0.916667	0.60800301	-1.508	0.1447
X2	1	3.916667	0.60800301	6.442	0.0001
X3	1	7.833333	0.60800301	12.884	0.0001

PREDICTION EQUATION: $Y = 80.535714 - 0.916667(x1) + 3.916667(x2) + 7.833333(x3)$

$R^2 = .8973$ and Adj $R^2 = .8845$ are both only slightly lower than 0.9. This model fits the data quite well. The F-test value, $F = 69.92$ is larger than $F(3, 24, .05) = 3.0$ with a small p-value = 0.0001. The T-test of $X1 = -1.508$ is not smaller than $-T(24, .025) = -2.064$. Hence X1 is not a significant regressor. The T-test of $X2 = 6.442$ is larger than $T(24, .025) = 2.064$. The T-test of $X3 = 12.884$ is also larger than $T(24, .025) = 2.064$ and X2, X3 have p-values less than 0.05. We can conclude that X2 and X3 have significant positive effect on the optimization of the yield.

6.19 Dependent Variable: Adsorption = adsorption index of phosphate
Independent Variables: Iron = amount of extractable iron
                        AL   = amount of extractable aluminum
                        PH   = pH of soil

Analysis of Variance

Source	DF	Sum of Squares	Mean Square	F Value	Prob>F
Model	3	3535.78331	1178.59444	57.057	0.0001
Error	9	185.90899	20.65655		
C Total	12	3721.69231			

Root MSE	4.54495	R-square	0.9500
Dep Mean	29.84615	Adj R-sq	0.9334
C.V.	15.22792		

Parameter Estimates

Variable	DF	Parameter Estimate	Standard Error	T for H0: Parameter=0	Prob > |T|
INTERCEP	1	-18.250879	20.74754742	-0.880	0.4019
Iron	1	0.112162	0.03083194	3.638	0.0054
AL	1	0.398339	0.11843926	3.363	0.0083
PH	1	1.420823	2.66300274	0.534	0.6066

Prediction Equation: Adsorption = −18.250879 + 0.112162(Iron) + 0.398339(AL) + 1.420823(PH)

$R^2 = .9500$ and Adj $R^2 = .9334$ are both larger than 0.9. This model fits the data very well.

The F-test value, F = 57.057 is larger than F(3,9, .05) = 3.86.
The T-test of iron, T = 3.638 is larger than T(9, .025) = 2.262.
The T-test of alum, T = 3.363 is also larger than T(9, .025) = 2.262.
However, the T-test of pH = 0.534 is smaller than T(9, .025) = 2.262.

The test of pH is the only one with p-value greater than 0.05. The pH is not a significant regressor. We can conclude that the amount of extractable iron and the amount of extractable aluminum have significant positive effect on the soils' adsorption of phosphate.

6.21 Dependent Variable: Date = first flight date
Independent Variables: Weight = specific power
                        Range = flight range
                        Payload = fraction of gross weight
                        Load = sustained load factor

The regression equation is
Date = −22.0 + 13.3 Weight + 31.0 Range − 77.0 Payload + 2.99 Load

Predictor	Coef	SE Coef	T	P
Constant	-22.05	52.31	-0.42	0.679
Weight	13.321	4.426	3.01	0.008
Range	31.02	11.73	2.65	0.017
Payload	-77.01	98.83	-0.78	0.447
Load	2.986	9.708	0.31	0.762

S = 34.1358 R-Sq = 70.9% R-Sq(adj) = 64.1%

Analysis of Variance

Source	DF	SS	MS	F	P
Regression	4	48279	12070	10.36	0.000
Residual Error	17	19809	1165		
Total	21	68088			

$R^2 = .709$ and Adj $R^2 = .641$ are both moderate values. This model fits the data moderately well. The F-test value, F = 10.36 with a p-value of 0.000. The T-test for weight (p-value = .008) and flight range (p-value = .017) show significance. We can conclude that the weight and the flight range have a positive significant effect on the first flight date.

6.23 Dependent Variable: Cost
Independent Variables: Files
                         Flows
                         Processes

The regression equation is
Cost = 1.96 + 0.12 Files + 0.177 Flows + 0.796 Processes

Predictor	Coef	SE Coef	T	P
Constant	1.962	5.608	0.35	0.737
Files	0.118	1.177	0.10	0.923
Flows	0.17673	0.07144	2.47	0.043
Processes	0.7964	0.2204	3.61	0.009

S = 9.91843   R-Sq = 96.1%   R-Sq(adj) = 94.5%

Analysis of Variance

Source	DF	SS	MS	F	P
Regression	3	17177.8	5725.9	58.21	0.000
Residual Error	7	688.6	98.4		
Total	10	17866.5			

$R^2 = .961$ and Adj $R^2 = .945$ are both high values. This model fits the data well. The F-test value, F = 58.21 with a p-value of 0.000. The T-test for flows (p-value = .043) and processes (p-value = .009) show significance. We can conclude that the flows and the processes have a positive significant effect on the cost of developing software.

6.25 Dependent Variable: Scorch Time
Independent Variables: Initial Temperature
                         Final Temperature
                         Open Mill Temperature

The regression equation is
Scorch Time = 11.1 + 0.0181 Initial Temp − 0.0585 Final Temp − .0718 Open Mill Temp

Predictor	Coef	SE Coef	T	P
Constant	11.116	1.361	8.17	0.000
Intial Temp	0.01805	0.01279	1.41	0.171
Final Temp	-0.05845	0.02539	-2.30	0.030
Open Mill Temp	-0.07176	0.01061	-6.76	0.000

S = 0.349073   R-Sq = 73.4%   R-Sq(adj) = 70.1%

Analysis of Variance

Source	DF	SS	MS	F	P
Regression	3	8.0639	2.6880	22.06	0.000
Residual Error	24	2.9245	0.1219		
Total	27	10.9883			

$R^2 = .734$ and Adj $R^2 = .701$ are both moderately high values. This model fits the data well. The F-test value, F = 22.06 with a p-value of 0.000. The T-test for final temperature (p-value = .030) and open mill temperature (p-value = .000) show significance. We can conclude that the final temperature and open mill temperature have negative significant effect on the scorch time.

6.27

Obs	Residual	Rstudent	Hat Diag H
1	0.1839	1.2617	0.2690
2	-0.0822	-0.5012	0.1812
3	0.1684	1.0404	0.1382
4	-0.1802	-1.1116	0.1237
5	-0.1496	-0.8908	0.0968
6	-0.1177	-0.6899	0.0944
7	0.2302	1.4339	0.0774
8	0.1676	0.9965	0.0771
9	-0.0940	-0.5443	0.0875
10	-0.2457	-1.5918	0.1137
11	-0.1255	-0.7679	0.1607
12	0.0821	0.5254	0.2555
13	0.1627	1.1477	0.3249

OUTLIERS: None because none of the |Rstudent| values is larger than 3.
LEVERAGE POINT: Obs 13 (Nickel = 0.819) is a potential leverage point because the Hat Diag value is larger than 2(2)/13 = .308

```
Stem-and-leaf of Rstudent   N  = 13
Leaf Unit = 0.10

    1    -1    5
    2    -1    1
   (5)   -0    87655
    6    -0
    6     0
    6     0    59
    4     1    0124
```

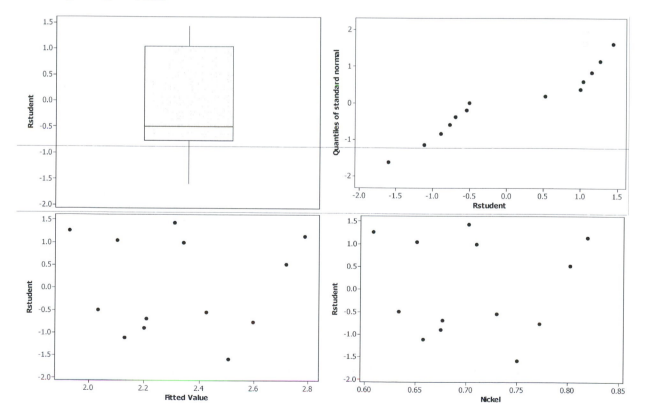

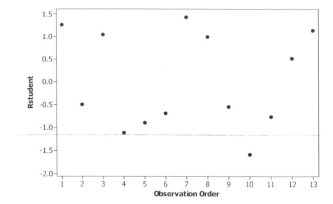

The stem-and-leaf plot, the boxplot, and the normal probability plot show some signs that the well-behaved distribution assumption may be violated. Other residual plots shows the constant variance assumption is not violated since the data are randomly scattered. We see a random scatter in the time-order plot.

6.29

Obs	Residual	Rstudent	Hat Diag
1	-0.0224	-0.58399	0.069216
2	-0.0095	-0.24496	0.065291
3	0.0040	0.10310	0.057650
4	-0.0041	-0.10556	0.055690
5	-0.0164	-0.42218	0.054548
6	-0.0092	-0.23577	0.052519
7	-0.0104	-0.26705	0.050232
8	0.0062	0.15908	0.049891
9	-0.0765	-2.09941	0.049722
10	0.0051	0.12972	0.044521
11	0.0284	0.73265	0.043376
12	0.0275	0.70792	0.040585
13	-0.0055	-0.14094	0.040585
14	-0.0152	-0.38991	0.040459
15	-0.0311	-0.80126	0.037241
16	0.0315	0.81149	0.037032
17	0.0491	1.28448	0.036128
18	0.0012	0.02966	0.034835
19	-0.0692	-1.85672	0.033526
20	0.0737	1.99085	0.030682
21	0.0610	1.61351	0.029417
22	0.0225	0.57366	0.029772
23	-0.0145	-0.36891	0.031500
24	-0.0032	-0.08187	0.033077
25	-0.0552	-1.45233	0.033077
26	0.0436	1.13998	0.045702
27	-0.0232	-0.60031	0.052690
28	0.0579	1.56025	0.070155
29	-0.0077	-0.19909	0.071363
30	0.0259	0.67830	0.071850
31	0.0695	1.94807	0.105953
32	-0.0414	-1.11757	0.107272
33	-0.0712	-2.08744	0.168868
34	-0.0212	-0.60734	0.225576

OUTLIERS: None because none of the |Rstudent| values is larger than 3.
LEVERAGE POINT: Observations 33 and 34 are potential leverage points since the Hat Diag values are larger than 2(2)/34 = .1176.

```
Stem-and-leaf of Rstudent   N  = 34
Leaf Unit = 0.10

    2    -2    00
    3    -1    8
    5    -1    41
    9    -0    8665
  (10)   -0    4332221110
   15     0    0111
   11     0    56778
    6     1    12
    4     1    5699
```

The stem-and-leaf plot, the boxplot, and the normal probability plot show no sign of grossly violated well-behaved distribution assumption. Other residual plots show the constant variance assumption should not be violated since the data are randomly scattered. Time-order plot indicates random scatter.

6.31

Obs	Residual	Rstudent	Hat Diag
1	5.50	1.12074	0.5
2	-5.50	-1.12074	0.5
3	4.50	0.85942	0.5
4	-4.50	-0.85942	0.5
5	-4.00	-0.74493	0.5
6	4.00	0.74493	0.5
7	-6.00	-1.27439	0.5
8	6.00	1.27439	0.5

```
Stem-and-leaf of Rstudent   N   = 8
Leaf Unit = 0.10

   2   -1   21
   4   -0   87
   4   -0
   4    0
   4    0   78
   2    1   12
```

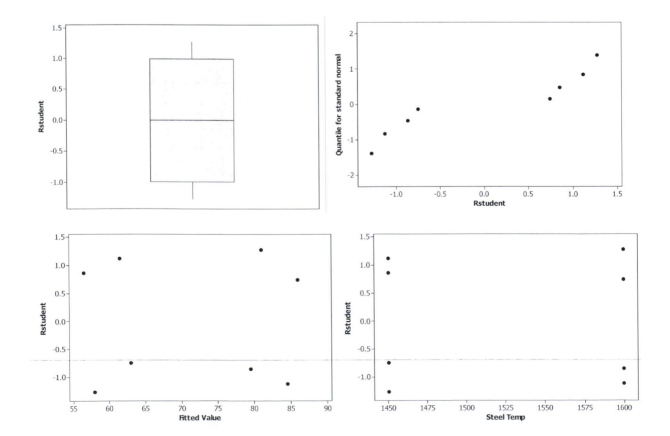

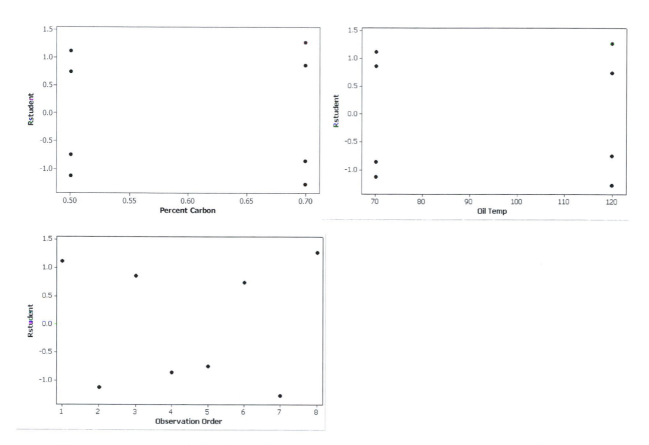

The stem-and-leaf plot, the boxplot, and the normal probability plot show no sign of grossly violated well-behaved distribution assumption. Time-order plot, residuals versus fitted values and residuals versus the regressors all look fine.

6.33	Obs	Residual	Rstudent	Hat Diag H
	1	1.7434	0.0898	0.2019
	2	1.1485	0.0578	0.1650
	3	-0.9365	-0.0459	0.1215
	4	-1.8264	-0.0910	0.1483
	5	-5.4164	-0.2827	0.2197
	6	42.7782	2.5453	0.1609
	7	35.2832	1.9235	0.1241
	8	-35.5017	-1.8814	0.0805
	9	-22.6916	-1.1490	0.1073
	10	-16.7816	-0.8715	0.1788
	11	-13.0827	-0.6693	0.1695
	12	-6.8777	-0.3408	0.1327
	13	6.4374	0.3111	0.0892
	14	20.0474	1.0113	0.1160
	15	23.8575	1.2775	0.1874
	16	-23.0153	-1.2303	0.1900
	17	-18.7103	-0.9616	0.1532
	18	-7.1952	-0.3520	0.1097
	19	7.2148	0.3584	0.1364
	20	13.5249	0.7097	0.2079

OUTLIERS: None because none of the |Rstudent| values is larger than 3.
LEVERAGE POINT: None because none of the Hat Diag values is larger than 2(3)/20 = .3.

```
Stem-and-leaf of Rstudent   N  = 20
Leaf Unit = 0.10

 1    -1    8
 3    -1    21
 6    -0    986
(5)   -0    33200
 9     0    0033
 5     0    7
 4     1    02
 2     1    9
 1     2
 1     2    5
```

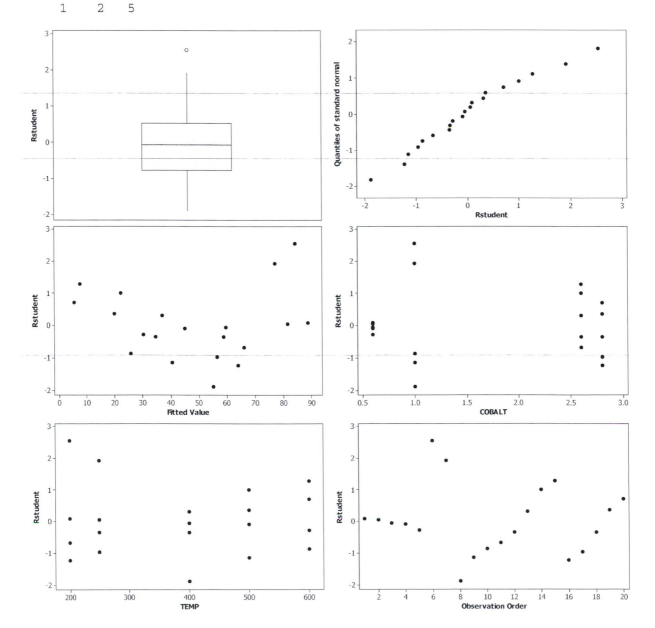

The stem-and-leaf plot, the boxplot, and the normal probability plot show no signs of grossly violated well-behaved distribution assumption. Residual plots exhibit curvature. There appears to be some sort of trending over time.

6.35

Obs	Residual	Rstudent	Hat Diag
1	-5.38	-0.18474	0.314414
2	-11.67	-0.39636	0.292752
3	-51.91	-1.80793	0.198058
4	-26.82	-0.91073	0.263315
5	-3.33	-0.10804	0.232468
6	19.36	0.60531	0.154615
7	-10.33	-0.34685	0.277635
8	-13.86	-0.41408	0.085310
9	20.29	0.61984	0.114101
10	-24.32	-0.73745	0.091877
11	25.17	0.89945	0.335292
12	17.51	0.52516	0.086267
13	-13.67	-0.41072	0.095905
14	33.24	1.05679	0.145144
15	-11.00	-0.33764	0.137090
16	42.95	1.59532	0.321566
17	-20.00	-0.66188	0.242535
18	-9.32	-0.33230	0.359828
19	-33.70	-1.10223	0.187812
20	-38.79	-1.32928	0.236343
21	41.35	1.55958	0.345899
22	74.22	4.30419	0.481776

OUTLIERS: One because observation 22 has a |Rstudent| value larger than 3.
LEVERAGE POINT: Observation 22 is a potential leverage point since the Hat Diag value is larger than $2(5)/22 = .4545$.

```
Stem-and-leaf of Rstudent   N  = 22
Leaf Unit = 0.10

  1    -1   8
  3    -1   31
  6    -0   976
 (8)   -0   44333311
  8     0
  8     0   5668
  4     1   0
  3     1   55
  1     2
  1     2
  1     3
  1     3
  1     4   3
```

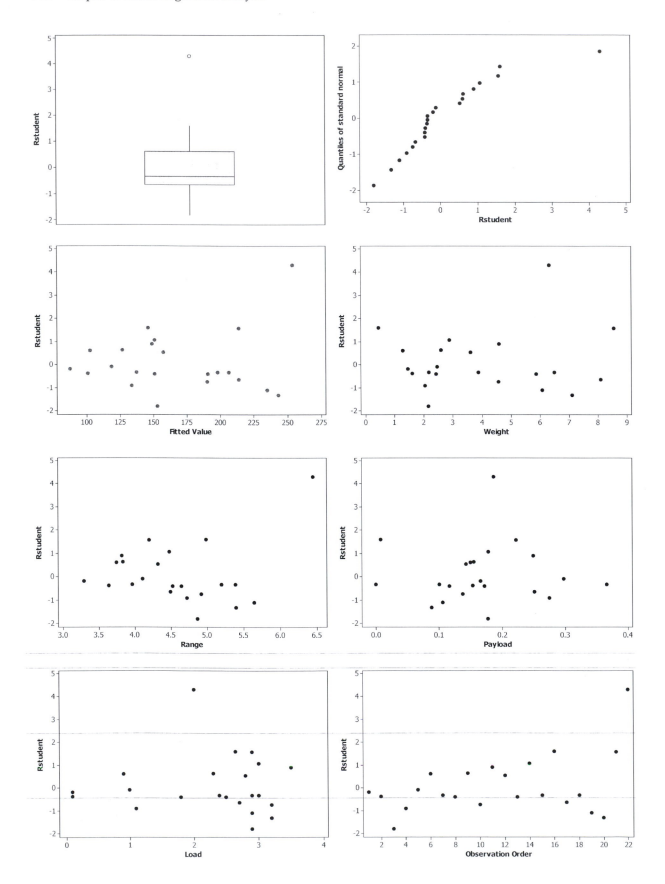

The stem-and-leaf plot, the boxplot, and the normal probability plot show no signs of grossly violated well-behaved distribution assumption. Other residual plots show the constant variance assumption should not be violated since the data are randomly scattered. An outlier is apparent in residual plots.

6.37	Obs.	Residual	Rstudent	Hat Diag
	1	0.037542	0.34271	0.563263
	2	−0.244023	−2.19115	0.135471
	3	0.117127	0.81128	0.161243
	4	−0.054480	−0.35248	0.129425
	5	−0.146261	−1.04352	0.149291
	6	0.044172	0.28801	0.149704
	7	0.170180	1.24537	0.130009
	8	−0.024549	−1.33207	0.983636
	9	0.100291	1.04027	0.597958

OUTLIERS: None because none have a |Rstudent| value larger than 3.
LEVERAGE POINT: Observation 8 is a potential leverage points since the Hat Diag Value is larger than $2(3)/9 = .667$.

```
Stem-and-leaf of Rstudent   N   = 9
Leaf Unit = 0.10

    1    -2    1
    1    -1
    3    -1    30
    3    -0
    4    -0    3
   (2)    0    23
    3     0    8
    2     1    02
```

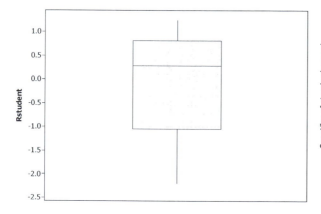

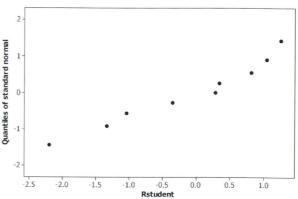

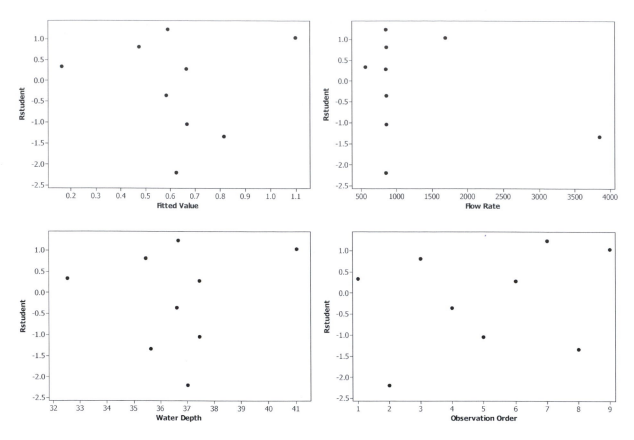

The stem-and-leaf plot, the boxplot, and the normal probability plot show no signs of grossly violated well-behaved distribution assumption. Other residual plots show the constant variance assumption should not be violated since the data are randomly scattered.

6.39 Dependent Variable: LOGVIS = log of viscosity
Independent Variable: TEMP = temperature

Analysis of Variance

Source	DF	Sum of Squares	Mean Square	F Value	Prob>F
Model	1	0.56388	0.56388	719.945	0.0001
Error	6	0.00470	0.00078		
C Total	7	0.56858			

Root MSE	0.02799	R-square	0.9917
Dep Mean	-0.31667	Adj R-sq	0.9904
C.V.	-8.83777		

Parameter Estimates

Variable	DF	Parameter Estimate	Standard Error	T for H0: Parameter=0	Prob > |T|
INTERCEP	1	0.376179	0.02765261	13.604	0.0001
TEMP	1	-0.011531	0.00042974	-26.832	0.0001

This model fits the data very well since the $R^2 = .9917$ and Adjusted $R^2 = .9904$ which are larger than .9. The F-Test value is 719.945 which is larger than $F(1,6, .05) = 5.99$. There should be at least one regressor is significant. The regressor Temp is significant since T-Test value $= -26.832$ which is smaller than $-T(6,.025) = -2.447$. Small p-value $= 0.0001$ for both tests.

| | | | Hat Diag |
Obs	Residual	Rstudent	H
1	0.0358	2.0964	0.4169
2	0.00433	0.1661	0.2734
3	-0.0172	-0.6452	0.1794
4	-0.0219	-0.8148	0.1309
5	-0.0214	-0.7963	0.1312
6	-0.0164	-0.6125	0.1789
7	-0.00626	-0.2411	0.2737
8	0.0431	3.2285	0.4157

OUTLIERS: Obs #8 because its |Rstudent| value is larger than 3.
LEVERAGE POINT: None because none of the Hat Diag values is larger than 2(2)/8= .5.

```
Stem-and-leaf of Rstudent   N  = 8
Leaf Unit = 0.10

   4  -0   8766
   4  -0   2
   3   0   1
   2   0
   2   1
   2   1
   2   2   0
   1   2
   1   3   2
```

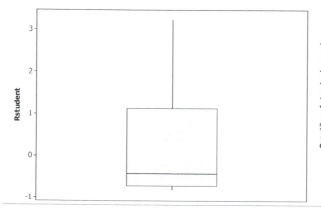
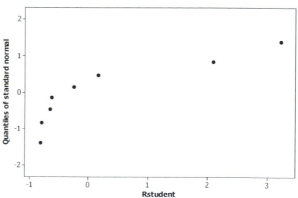

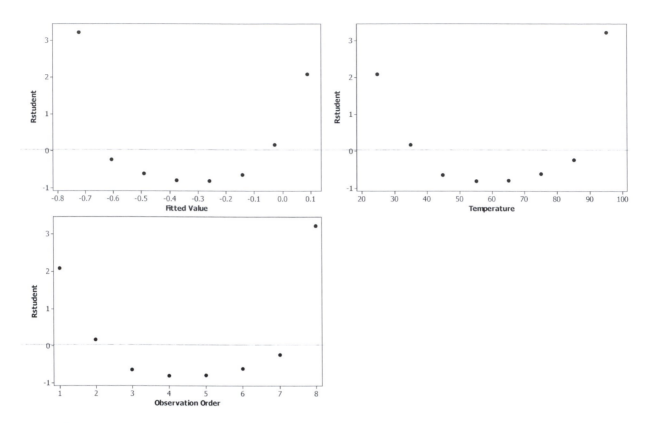

The stem-and-leaf plot, the boxplot, and the normal probability plot show that the distribution is skewed to the right. The well-behaved distribution assumption may be violated. Other residual plots show the constant variance assumption may be violated since the data shows a pattern. Curvature is apparent, so we may need to consider quadratic terms.

6.41

Variable	DF	Variance Inflation
INTERCEP	1	0.00000000
TEMP	1	1.33108603
CONC	1	1.33108603

Correlation of Estimates

CORRB	INTERCEP	TEMP	CONC
INTERCEP	1.0000	-0.8506	-0.8800
TEMP	-0.8506	1.0000	0.4987
CONC	-0.8800	0.4987	1.0000

Collinearity Diagnostics

Number	Eigenvalue	Condition Index	Var Prop INTERCEP	Var Prop TEMP	Var Prop CONC
1	2.99980	1.00000	0.0000	0.0000	0.0000
2	0.0001805	128.91721	0.0002	0.2896	0.2136
3	0.0000199	388.08126	0.9998	0.7104	0.7864

Temp and Conc. have a correlation of .4987. There are also problems with condition numbers; consider more experimentation.

6.43

Variable	DF	Variance Inflation
INTERCEP	1	0.00000000
COBALT	1	1.00000000
TEMP	1	1.00000000

Correlation of Estimates

CORRB	INTERCEP	COBALT	TEMP
INTERCEP	1.0000	-0.5456	-0.7824
COBALT	-0.5456	1.0000	0.0000
TEMP	-0.7824	0.0000	1.0000

Collinearity Diagnostics

Number	Eigenvalue	Condition Index	Var Prop INTERCEP	Var Prop COBALT	Var Prop TEMP
1	2.75261	1.00000	0.0114	0.0269	0.0156
2	0.19048	3.80147	0.0189	0.7714	0.2210
3	0.05691	6.95455	0.9697	0.2017	0.7635

There are no collinearity problems among predictors; designed experiment.

6.45

Variable	DF	Variance Inflation
INTERCEP	1	0.00000000
VELO	1	1.10931510
VISC	1	1.10583953
SIZE	1	1.00434606

Correlation of Estimates

CORRB	INTERCEP	VELO	VISC	SIZE
INTERCEP	1.0000	-0.5324	-0.2237	-0.8561
VELO	-0.5324	1.0000	0.3079	0.0574
VISC	-0.2237	0.3079	1.0000	-0.0128
SIZE	-0.8561	0.0574	-0.0128	1.0000

Collinearity Diagnostics

Number	Eigenvalue	Condition Index	Var Prop INTERCEP	Var Prop VELO	Var Prop VISC	Var Prop SIZE
1	3.17818	1.00000	0.0024	0.0092	0.0248	0.0033
2	0.72823	2.08907	0.0007	0.0131	0.8146	0.0009
3	0.07842	6.36607	0.0243	0.8136	0.1377	0.1322
4	0.01517	14.47598	0.9726	0.1640	0.0229	0.8636

There is no collinearity problem in this model since all VIF <5 and all Condition Index < 30.

6.47

Variable	DF	VIF
Intercept	1	0.000
Carbon	1	1.164
Ash	1	1.040
Sulfur	1	1.178

Collinearity Diagnostics

Number	Eigenvalue	Condition Index
1	3.8775	1.00
2	0.0763	7.13
3	0.0405	9.78
4	0.0057	26.08

There is no collinearity problem in this model since all VIF < 5 and all Condition Index < 30.

6.49

Variable	DF	VIF
Intercept	1	0.000
Initial Temp	1	3.526
Final Temp	1	3.629
Open Mill Temp	1	1.101

Collinearity Diagnostics

Number	Eigenvalue	Condition Index
1	3.9839	1.00
2	0.0100	19.96
3	0.0055	80.54
4	0.0007	75.44

There is a collinearity problem since two of the condition indices are greater than 30.

CHAPTER 7

INTRODUCTION TO 2^k FACTORIAL EXPERIMENTS

7.1

Pressure	Ratio	Uniformity
−1	−1	(1) 8.6
1	−1	a 3.4
−1	1	b 6.9
1	1	ab 5.1

Main effect: pressure = $(5.1 + 3.4 − 6.9 − 8.6)/2 = −3.5$
Main effect: ratio = $(5.1 − 3.4 + 6.9 − 8.6)/2 = 0$
Interaction: pressure and ratio = $(5.1 − 3.4 − 6.9 + 8.6)/2 = 1.7$

7.3

Location	Length	Failure Load
−1	−1	(1) 342
1	−1	a 378
−1	1	b 463
1	1	ab 517

Main effect: location = $(378 + 517 − 342 − 463)/2 = 45$
Main effect: length = $(463 + 517 − 342 − 378)/2 = 130$
Interaction: location and length = $(342 + 517 − 378 − 463)/2 = 9$

7.5 a

Angle	Height	Distance
−1	−1	(1) 27.0
1	−1	a 74.0
−1	1	b 64.5
1	1	ab 147.5

Main effect: Angle = $(147.5 + 74 − 64.5 − 27)/2 = 65$
Main effect: Height = $(147.5 − 74 + 64.5 − 27)/2 = 55.5$
Interaction: Angle and Height = $(147.5 − 74 − 64.5 + 27)/2 = 18$

b analysis used design variables

Dependent Variable: DISTANCE
Independent Variables: ANGLE; HEIGHT

Analysis of Variance

Source	DF	Sum of Squares	Mean Square	F Value	Prob>F
Model	3	15258.50000	5086.16667	61.464	0.0008
Error	4	331.00000	82.75000		
C Total	7	15589.50000			

R-square	0.9788	overall model significant
Adj R-sq	0.9628	R-sq 97.88% excellent

Parameter Estimates

Variable	DF	Parameter Estimate	Standard Error	T for H0: Parameter=0	Prob > |T|
INTERCEP	1	78.250000	3.21617008	24.330	0.0001
ANGLE	1	32.500000	3.21617008	10.105	0.0005
HEIGHT	1	27.750000	3.21617008	8.628	0.0010
INTERACT	1	9.000000	3.21617008	2.798	0.0489

all factors important at .05 level: we have positive main effects and positive interaction

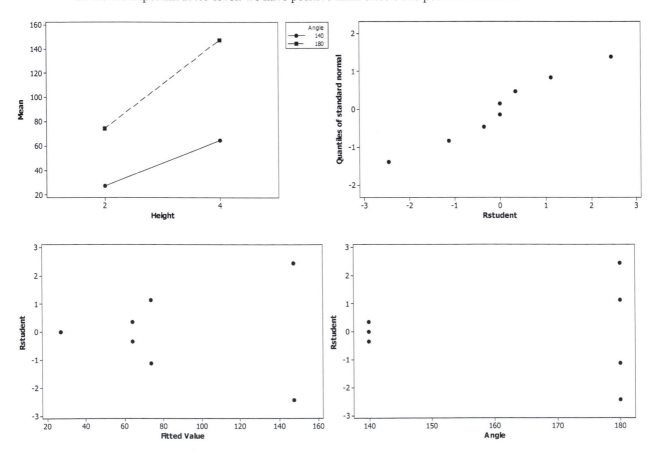

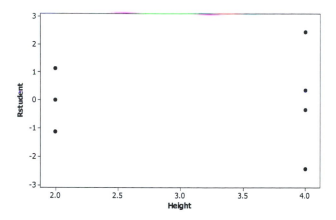

The residual plots indicate that the variance may not be constant.
Normal probability plot gives straight line.

7.7   **a**

Mixture	Temperature		Toughness
−1	−1	(1)	15.445
1	−1	a	13.760
−1	1	b	13.620
1	1	ab	15.740

Main effect: mixture = (15.74 +13.76 −13.62 −15.445)/2 = .2175
Main effect: temperature = (15.74 + 13.62 − 13.76 − 15.445)/2= .0775
Interaction: mixture and temperature = (15.74 + 15.445 − 13.62 − 13.76)/2=1.9025

Estimated Effects and Coefficients for Toughness (coded units)

Term	Effect	Coef	SE Coef	T	P
Constant		14.6413	0.1254	116.75	0.000
Mixture	0.2175	0.1087	0.1254	0.87	0.435
Temp	0.0775	0.0387	0.1254	0.31	0.773
Mixture*Temp	1.9025	0.9512	0.1254	7.59	0.002

S = 0.354701   R-Sq = 93.59%   R-Sq(adj) = 88.78%

Analysis of Variance for Toughness (coded units)

Source	DF	Seq SS	Adj SS	Adj MS	F	P
Main Effects	2	0.10662	0.10662	0.05331	0.42	0.681
2-Way Interactions	1	7.23901	7.23901	7.23901	57.54	0.002
Residual Error	4	0.50325	0.50325	0.12581		
Pure Error	4	0.50325	0.50325	0.12581		
Total	7	7.84889				

There is a significant interaction between the asphalt mixture and the temperature, p-value=.002.

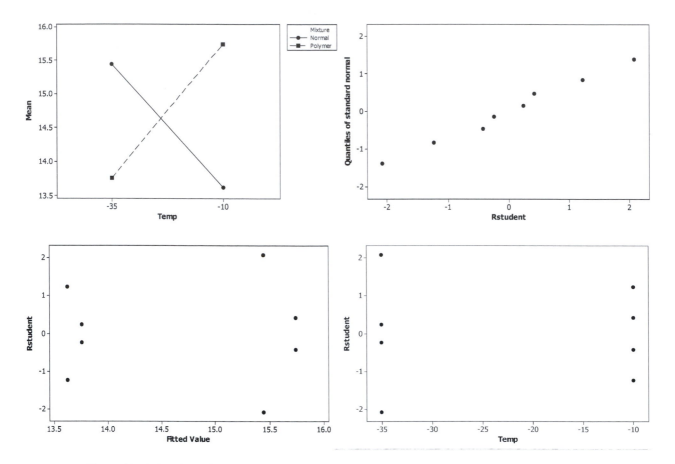

The residual plots look fine.

7.9   a,b,c   Factorial design 2^4 - design variables (natural units)
            Gas Input: B = Bottom and T = Top ; Temp. Gradient: F = Flat and R = Ramp

Random Order	Pressure (Torr)	Temp. Gradient	Furnace Temp. (°C)	Gas Input
3	−1 (1)	−1 (F)	−1 (625°)	−1 (B)
12	1 (3)	−1 (F)	−1 (625°)	−1 (B)
16	−1 (1)	1 (R)	−1 (625°)	−1 (B)
11	1 (3)	1 (R)	−1 (625°)	−1 (B)
14	−1 (1)	−1 (F)	1 (725°)	−1 (B)
10	1 (3)	−1 (F)	1 (725°)	−1 (B)
6	−1 (1)	1 (R)	1 (725°)	−1 (B)
9	1 (3)	1 (R)	1 (725°)	−1 (B)
5	−1 (1)	−1 (F)	−1 (625°)	1 (T)
4	1 (3)	−1 (F)	−1 (625°)	1 (T)
8	−1 (1)	1 (R)	−1 (625°)	1 (T)
13	1 (3)	1 (R)	−1 (625°)	1 (T)
7	−1 (1)	−1 (F)	1 (725°)	1 (T)
2	1 (3)	−1 (F)	1 (725°)	1 (T)
15	−1 (1)	1 (R)	1 (725°)	1 (T)
1	1 (3)	1 (R)	1 (725°)	1 (T)

7.11 a,b,c	Random Order	Rate	Intensity	Air	Conc.	Time
	13	−1 (−0.2)	−1 (160)	−1 (yes)	−1 (7.5)	−1 (32.5)
	27	1 (−0.1)	−1 (160)	−1 (yes)	−1 (7.5)	−1 (32.5)
	23	−1 (−0.2)	1 (200)	−1 (yes)	−1 (7.5)	−1 (32.5)
	5	1 (−0.1)	1 (200)	−1 (yes)	−1 (7.5)	−1 (32.5)
	22	−1 (−0.2)	−1 (160)	1 (no)	−1 (7.5)	−1 (32.5)
	31	1 (−0.1)	−1 (160)	1 (no)	−1 (7.5)	−1 (32.5)
	32	−1 (−0.2)	1 (200)	1 (no)	−1 (7.5)	−1 (32.5)
	10	1 (−0.1)	1 (200)	1 (no)	−1 (7.5)	−1 (32.5)
	16	−1 (−0.2)	−1 (160)	−1 (yes)	1 (12.5)	−1 (32.5)
	4	1 (−0.1)	−1 (160)	−1 (yes)	1 (12.5)	−1 (32.5)
	24	−1 (−0.2)	1 (200)	−1 (yes)	1 (12.5)	−1 (32.5)
	17	1 (−0.1)	1 (200)	−1 (yes)	1 (12.5)	−1 (32.5)
	26	−1 (−0.2)	−1 (160)	1 (no)	1 (12.5)	−1 (32.5)
	29	1 (−0.1)	−1 (160)	1 (no)	1 (12.5)	−1 (32.5)
	30	−1 (−0.2)	1 (200)	1 (no)	1 (12.5)	−1 (32.5)
	9	1 (−0.1)	1 (200)	1 (no)	1 (12.5)	−1 (32.5)
	3	−1 (−0.2)	−1 (160)	−1 (yes)	−1 (7.5)	1 (57.5)
	1	1 (−0.1)	−1 (160)	−1 (yes)	−1 (7.5)	1 (57.5)
	15	−1 (−0.2)	1 (200)	−1 (yes)	−1 (7.5)	1 (57.5)
	6	1 (−0.1)	1 (200)	−1 (yes)	−1 (7.5)	1 (57.5)
	11	−1 (−0.2)	−1 (160)	1 (no)	−1 (7.5)	1 (57.5)
	18	1 (−0.1)	−1 (160)	1 (no)	−1 (7.5)	1 (57.5)
	12	−1 (−0.2)	1 (200)	1 (no)	−1 (7.5)	1 (57.5)
	25	1 (−0.1)	1 (200)	1 (no)	−1 (7.5)	1 (57.5)
	20	−1 (−0.2)	−1 (160)	−1 (yes)	1 (12.5)	1 (57.5)
	28	1 (−0.1)	−1 (160)	−1 (yes)	1 (12.5)	1 (57.5)
	14	−1 (−0.2)	1 (200)	−1 (yes)	1 (12.5)	1 (57.5)
	21	1 (−0.1)	1 (200)	−1 (yes)	1 (12.5)	1 (57.5)
	7	−1 (−0.2)	−1 (160)	1 (no)	1 (12.5)	1 (57.5)
	2	1 (−0.1)	−1 (160)	1 (no)	1 (12.5)	1 (57.5)
	19	−1 (−0.2)	1 (200)	1 (no)	1 (12.5)	1 (57.5)
	8	1 (−0.1)	1 (200)	1 (no)	1 (12.5)	1 (57.5)

7.13 a	Angle	Height	Stop	Hook	Distance
	180	3	3	3	(1) 363.5
	Full	3	3	3	a 403.5
	180	4	3	3	b 438.0
	Full	4	3	3	ab 480.0
	180	3	5	3	c 381.5
	Full	3	5	3	ac 438.5
	180	4	5	3	bc 475.5
	Full	4	5	3	abc 545.0
	180	3	3	5	d 419.5
	Full	3	3	5	ad 484.0
	180	4	3	5	bd 491.0
	Full	4	3	5	abd 537.5
	180	3	5	5	cd 456.5
	Full	3	5	5	acd 531.0
	180	4	5	5	bcd 482.5
	Full	4	5	5	abcd 585.5

```
Estimated Effects and Coefficients for Distance (coded units)

Term                      Effect     Coef  SE Coef        T      P
Constant                           469.563    2.683   174.98  0.000
Angle                     62.125   31.063    2.683    11.58  0.000
Height                    69.625   34.812    2.683    12.97  0.000
Stop                      34.875   17.438    2.683     6.50  0.000
Hook                      57.750   28.875    2.683    10.76  0.000
Angle*Height               3.125    1.562    2.683     0.58  0.569
Angle*Stop                13.875    6.937    2.683     2.59  0.020
Angle*Hook                10.000    5.000    2.683     1.86  0.081
Height*Stop                0.625    0.313    2.683     0.12  0.909
Height*Hook              -18.250   -9.125    2.683    -3.40  0.004
Stop*Hook                 -4.000   -2.000    2.683    -0.75  0.467
Angle*Height*Stop          7.125    3.563    2.683     1.33  0.203
Angle*Height*Hook         -0.500   -0.250    2.683    -0.09  0.927
Angle*Stop*Hook            2.750    1.375    2.683     0.51  0.615
Height*Stop*Hook         -11.750   -5.875    2.683    -2.19  0.044
Angle*Height*Stop*Hook     4.500    2.250    2.683     0.84  0.414


S = 15.1802     PRESS = 14748
R-Sq = 96.84%   R-Sq(pred) = 87.36%   R-Sq(adj) = 93.88%


Analysis of Variance for Distance (coded units)

Source                DF  Seq SS  Adj SS   Adj MS       F      P
Main Effects           4  106068  106068  26517.0  115.07  0.000
2-Way Interactions     6    5214    5214    869.0    3.77  0.016
3-Way Interactions     4    1573    1573    393.3    1.71  0.198
4-Way Interactions     1     162     162    162.0    0.70  0.414
Residual Error        16    3687    3687    230.4
  Pure Error          16    3687    3687    230.4
Total                 31  116704
```

b Normal probability plot of the effects:

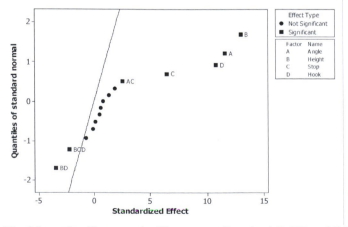

c All of the main effects are significant as well as the AC, BD and BCD interactions.

7.15 a

Factor	Effect
Battery	437.50
Connector	−28.50
Temp	−20.00
Battery*Connector	−30.50
Battery*Temp	−40.00
Connector*Temp	31.00
Battery*Connector*Temp	32.00

b

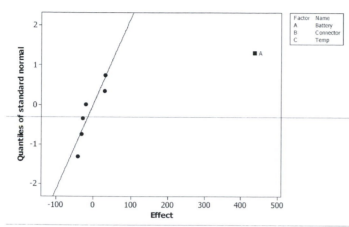

c Using high-quality batteries seems to be the most important factor as related to time to discharge.

7.17 a

Main Effects	Two-Way Interaction	Three-Way Interaction	Four-Way Interaction
A glue .62	AB −.009	ABC −.28	ABCD −.03
B predrying temp. 1.09	AC .32	ABD −.02	
C tunnel temp. .69	AD −.099	ACD .39	
D pressure −.19	BC .299	BCD −.04	
	BD .75		
	CD .03		

b

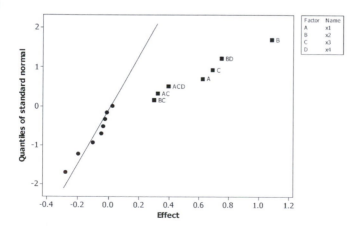

c Main effects A, B, C appear important. However Factor D appears important through interactions with other factors.

7.19 a,b,c   Fractional factorial design 2^{5-2}_{III} - design variables (natural units)

Random Order	A Pressure (bar)	B Temp. °C	C Moisture (% by weight)	D Flow Rate (L/min)	E Particle Size (mm)
3	−1 (415)	−1 (25)	−1 (5)	1 (60)	1 (4.05)
7	1 (550)	−1 (25)	−1 (5)	−1 (40)	1 (1.28)
1	−1 (415)	1 (95)	−1 (5)	−1 (40)	1 (4.05)
2	1 (550)	1 (95)	−1 (5)	1 (60)	−1 (1.28)
6	−1 (415)	−1 (25)	−1 (15)	1 (60)	−1 (1.28)
5	1 (550)	−1 (25)	−1 (15)	−1 (40)	1 (4.05)
8	−1 (415)	−1 (95)	−1 (15)	−1 (40)	−1 (1.28)
4	1 (550)	1 (95)	1 (15)	1 (60)	1 (4.00)

Alias Structure

I	=	ABD	=	ACE	=	BCDE
A	=	BD	=	CE	=	ABCDE
B	=	AD	=	ABCE	=	CDE
C	=	ABCD	=	AE	=	BDE
D	=	AB	=	ACDE	=	BCE
E	=	ABDE	=	AC	=	BCD
CD	=	ABC	=	ADE	=	BE

7.21 a,b,c   Fractional factorial design 2^{4-1}_{IV}

Random Order	A Angle	B Height	C Stop	D Hook
4	Full	4	3	3
2	Full	3	3	5
1	180	3	3	3
5	180	3	5	5
3	180	4	3	5
7	180	4	5	3
8	Full	4	5	5
6	Full	3	5	3

Alias Structure
```
I  = ABCD
A  = BCD
B  = ACD
C  = ABD
D  = ABC
AB = CD
AC = BD
AD = BC
```

7.23 a 2^{5-2}_{III}: 1/4 fraction of 2^5, Resolution III

b I =BCDE = −ACD = −ABE
A=ABCDE= −CD = −BE
B=CDE = −ABCD = −AE
C=BDE = −AD = −ABCE
D=BCE = −AC = −ABDE
E=BCD = −ACDE = −AB
BC = DE = −ABD = −ACE
BD = CE = −ABC = −ADE

c

Source	Effect
A	.04875
B	.04625
C	.00875
D	−.12325
E	−.12125
BC	−.04375
BD	−.00625

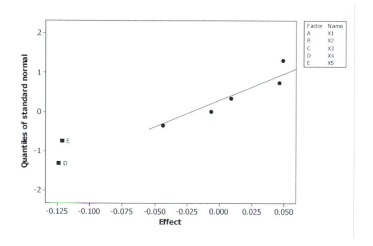

Dependent Variable: Y = carbon in cast iron
Independent Variables: D = oxygen flow; E = muffle temp.

Analysis of Variance

Source	DF	Sum of Squares	Mean Square	F Value	Prob>F
Model	2	0.06003	0.03002	11.46	0.014
Error	5	0.01309	0.00262		
C Total	7	0.07312			

R-square	0.8210	overall model significant	
Adj R-sq	0.7494	R-sq 82.1% fairly good	

Parameter Estimates

Variable	DF	Parameter Estimate	Standard Error	T for H0: Parameter=0	Prob > |T|
INTERCEP	1	3.084375	0.01809	170.50	0.000
D	1	−0.061875	0.01809	−3.42	0.019
E	1	−0.060625	0.01809	−3.35	0.020

Oxygen flow and Muffle Temperature appear important.

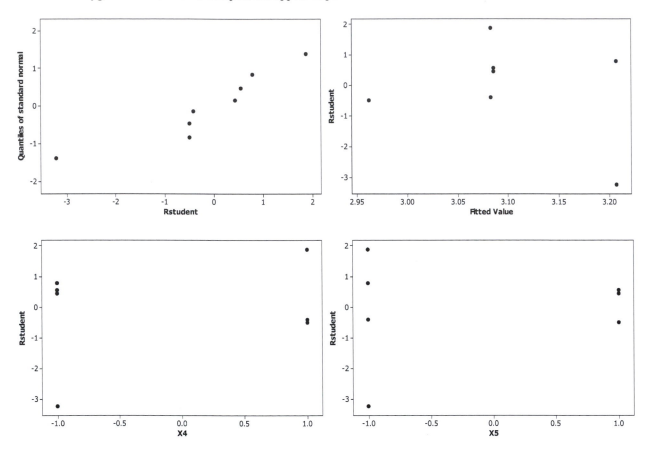

No major problems from the residual plots.
Normal probability plot gives reason for concern about normality assumption. However, there are only eight observations.

7.25 a 2^{6-3}_{III}: 1/8 Fraction of 2^6, Resolution III

    b I = BCDE = −ADE = −ABC = −BDF = −CEF = ABEF = ACDF
       with x_1 = A, x_2 = B, x_3 = C, x_4 = D, x_5 = E, x_6 = F

    c

Source	Effect
A	−.08975
B	.00650
C	−.04525
D	−.02600
E	−.04175
F	.01950
AF	−.00775

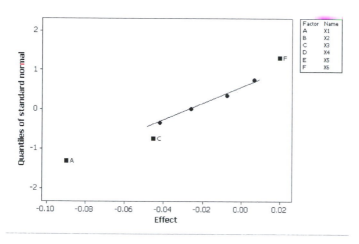

Dependent Variable: Y = manganese in cast iron
Independent Variables: A = titration speed; C = AgNO$_3$ addition; F = sodium arsenite

Analysis of Variance

Source	DF	Sum of Squares	Mean Square	F Value	Prob>F
Model	3	0.02097	0.00699	5.543	0.0658
Error	4	0.00504	0.00126		
C Total	7	0.02601			

R-square	0.8061	overall model significant at .10 level	
Adj R-sq	0.6607	R-sq 80.61% okay	

Parameter Estimates

Variable	DF	Parameter Estimate	Standard Error	T for H0: Parameter=0	Prob > |T|
INTERCEP	1	0.090000	0.01255332	7.169	0.0020
A	1	-0.044875	0.01255332	-3.575	0.0233
C	1	-0.022625	0.01255332	-1.802	0.1458
F	1	0.009750	0.01255332	0.777	0.4807

A = titration speed appears important; p-value < .05

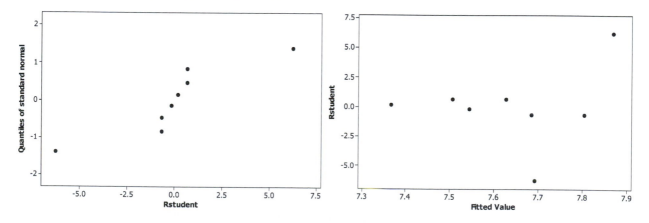

Residual plots show possible violations of assumptions.

7.27 a 2^{4-1}_{IV} fractional factorial

b <u>Alias Structure</u>
```
I  = ABCD
A  = BCD
B  = ACD
C  = ABD
D  = ABC
AB = CD
AC = BD
AD = BC
```

c

Term	Effect
pH	5.25
salt	-36.25
time	-1.25
temp	17.25
pH*salt	24.75
pH*time	-2.25
pH*temp	1.25

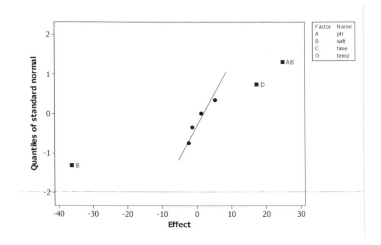

Estimated Effects and Coefficients for stability (coded units)

Term	Effect	Coef	SE Coef	T	P
Constant		70.63	0.8260	85.50	0.000
pH	5.25	2.62	0.8260	3.18	0.050
salt	-36.25	-18.12	0.8260	-21.94	0.000
temp	17.25	8.63	0.8260	10.44	0.002
pH*salt	24.75	12.38	0.8260	14.98	0.001

S = 2.33631   R-Sq = 99.64%   R-Sq(adj) = 99.15%
Analysis of Variance for stability (coded units)

Source	DF	Seq SS	Adj SS	Adj MS	F	P
Main Effects	3	3278.37	3278.37	1092.79	200.21	0.001
2-Way Interactions	1	1225.13	1225.13	1225.13	224.45	0.001
Residual Error	3	16.38	16.38	5.46		
Total	7	4519.87				

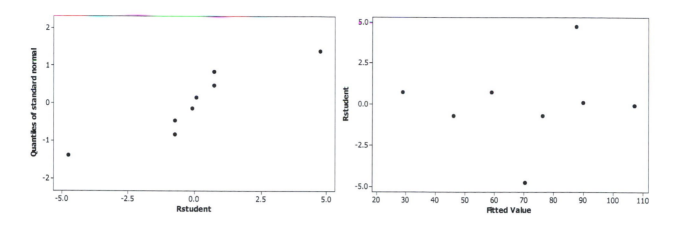

7.29 a 2^{7-4}_{III} fractional factorial

b <u>Alias Structure</u>
I = ABD = ACE = AFG = BCF = BEG = CDG = DEF = ABCG = ABEF = ACDF = ADEG =
   BCDE = BDFG = CEFG = ABCDEFG
A = BD = CE = FG = BCG = BEF = CDF = DEG = ABCF = ABEG = ACDG = ADEF =
   ABCDE = ABDFG = ACEFG = BCDEFG
B = AD = CF = EG = ACG = AEF = CDE = DFG = ABCE = ABFG = BCDG = BDEF =
   ABCDF = ABDEG = BCEFG = ACDEFG
C = AE = BF = DG = ABG = ADF = BDE = EFG = ABCD = ACFG = BCEG = CDEF =
   ABCEF = ACDEG = BCDFG = ABDEFG
D = AB = CG = EF = ACF = AEG = BCE = BFG = ACDE = ADFG = BCDF = BDEG =
   ABCDG = ABDEF = CDEFG = ABCEFG
E = AC = BG = DF = ABF = ADG = BCD = CFG = ABDE = AEFG = BCEF = CDEG =
   ABCEG = ACDEF = BDEFG = ABCDFG
F = AG = BC = DE = ABE = ACD = BDG = CEG = ABDF = ACEF = BEFG = CDFG =
   ABCFG = ADEFG = BCDEF = ABCDEG
G = AF = BE = CD = ABC = ADE = BDF = CEF = ABDG = ACEG = BCFG = DEFG =
   ABEFG = ACDFG = BCDEG = ABCDEF

c Estimated Effects and Coefficients for force (coded units)

Term	Effect	Coef	SE Coef	T	P
Constant		58.078	0.8258	70.32	0.000
Inj Sepped	22.770	11.385	0.8258	13.79	0.000
Mold Temp	-8.682	-4.341	0.8258	-5.26	0.000
Melt temp	7.708	3.854	0.8258	4.67	0.000
Cooling time	-4.475	-2.238	0.8258	-2.71	0.014
Ejection speed	5.457	2.728	0.8258	3.30	0.004

S = 4.04580 R-Sq = 93.47% R-Sq(adj) = 91.66%

Analysis of Variance for force (coded units)

Source	DF	Seq SS	Adj SS	Adj MS	F	P
Main Effects	5	4218.38	4218.38	843.68	51.54	0.000
Residual Error	18	294.63	294.63	16.37		
Lack of Fit	2	74.73	74.73	37.36	2.72	0.096
Pure Error	16	219.90	219.90	13.74		
Total	23	4513.01				

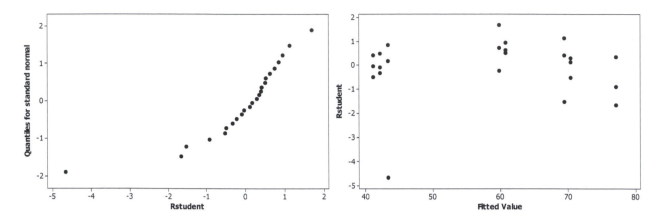

There is a severe outlier, however it does not affect the overall results.

d They could have run a 2^{7-5}_{IV} fractional factorial. That would be 16 runs.

With the other 8 runs they could: 1) replicate 8 of the 16 runs, 2) add 8 center runs, or 3) add 8 runs from the ones not run to break some of the aliasing.

CHAPTER 8

INTRODUCTION TO RESPONSE SURFACE METHODOLOGY

8.1   Fitted equation: uniformity = 6 + −1.75 Pressure

	Pressure
b_j	−1.75
c_j	6.00
d_j	27.5

Path of Steepest descent

	(Design Variables)	(Natural Units)
Run	Ratio	Ratio
1	1.0	70
2	1.1	72.75
3	1.2	75.5
4	1.3	78.25
5	1.4	81

8.3   Fitted equation: Failure Load = 425 + 22.5 Location + 65 Length

	Location x_1	Length x_2
b_j	22.5	65
c_j	800.0	240
d_j	400.0	80

$x_1 = (22.5/65)\, x_2 = .3462x_2$

Path of Steepest Ascent

	Design Variables		Natural Units	
Run	x_1	x_2	Location	Length
1	.3462	1.0	938.5	320
2	.3808	1.1	952.3	328
3	.4154	1.2	966.2	336
4	.4501	1.3	980.1	344
5	.4847	1.4	993.9	352

8.5 Fitted first order model: Thickness = .6725 − .03625 glue + .07 temp. + .065 pressure

	Glue x_1	Temp. x_2	Pressure x_3
b_j	−0.03625	.07	.065
c_j	?	?	?
d_j	?	?	?

$x_1 = (−.03625/.07) x_2 = −.518 x_2$ $x_3 = (.065/.07) x_2 = .93 x_2$

Path of steepest descent
      (Design Variables)

Run	x_1	x_2	x_3
1	.518	−1.0	−0.93
2	.570	−1.1	−1.02
3	.622	−1.2	−1.11
4	.673	−1.3	−1.21
5	.725	−1.4	−1.30

8.7 First order model suggested by 7.18: $y = 8.0 − 1.0 x_4 − 3.75 x_6$

	x_1	x_6
b_j	−1.0	−3.75
c_j	4.25	9.65
d_j	2.75	.1

$x_1 = (−1/−3.75) x_6 = .267 x_6$

Path of steepest ascent

Run	(Design Variables) x_1	x_6	(Natural Units) x_1	x_6
1	−.267	−1.0	3.52	9.55
2	−.293	−1.1	3.44	9.54
3	−.320	−1.2	3.37	9.53
4	−.347	−1.3	3.30	9.52
5	−.373	−1.4	3.22	9.51

8.9 **a** Face-centered cube ccd $\alpha = 1$ &

Factorial portion 2^{5-1}_V: resolution V with I = ABCDE, 10 axial runs, and one center run

Pressure (bar)		Temp. (°C)		Moisture (% by weight)		Flow Rate (L/min)		Particle Size (mm)	
−1	(415)	−1	(25)	−1	(5)	−1	(40)	1	(4.05)
1	(550)	−1	(25)	−1	(5)	−1	(40)	−1	(1.28)
−1	(415)	1	(95)	−1	(5)	−1	(40)	−1	1.24
1	(550)	1	(95)	−1	(5)	−1	(40)	1	(4.05)
−1	(415)	−1	(25)	1	(15)	−1	(40)	−1	(1.28)
1	(550)	−1	(25)	1	(15)	−1	(40)	1	(4.05)
−1	(415)	1	(95)	1	(15)	−1	(40)	1	(4.05)
1	(550)	1	(95)	1	(15)	−1	(40)	−1	(1.28)
−1	(415)	−1	(25)	−1	(5)	1	(60)	−1	(1.28)
1	(550)	−1	(25)	−1	(5)	1	(60)	1	(4.05)
−1	(415)	1	(95)	−1	(5)	1	(60)	1	(4.05)
1	(550)	1	(95)	−1	(5)	1	(60)	−1	(1.28)
−1	(415)	−1	(25)	1	(15)	1	(60)	1	(4.05)
1	(550)	−1	(25)	1	(15)	1	(60)	−1	(1.28)
−1	(415)	1	(95)	1	(15)	1	(60)	−1	(1.28)
1	(550)	1	(95)	1	(15)	1	(60)	1	(4.05)
−1	(415)	0	(60)	0	(10)	0	(50)	0	(2.665)
1	(550)	0	(60)	0	(10)	0	(50)	0	(2.665)
0	(482.5)	−1	(25)	0	(10)	0	(50)	0	(2.665)
0	(482.5)	1	(95)	0	(10)	0	(50)	0	(2.665)
0	(482.5)	0	(60)	−1	(5)	0	(50)	0	(2.665)
0	(482.5)	0	(60)	1	(15)	0	(50)	0	(2.665)
0	(482.5)	0	(60)	0	(10)	−1	(40)	0	(2.665)
0	(482.5)	0	(60)	0	(10)	1	(60)	0	(2.665)
0	(482.5)	0	(60)	0	(10)	0	(50)	−1	(1.28)
0	(482.5)	0	(60)	0	(10)	0	(50)	1	(4.05)
0	(482.5)	0	(60)	0	(10)	0	(50)	0	(2.665)

b rotatable ccd $\alpha = 16^{.25} = 2$
  Factorial portion 2^{5-1}_V: resolution V with I = ABCDE, 10 axial runs, and one center run
  Note this design uses negative temperature (possibly bad) and particle size (impossible)!

Pressure (bar)		Temp. (°C)		Moisture (% by weight)		Flow Rate (L/min)		Particle Size (mm)	
−1	(415)	−1	(25)	−1	(5)	−1	(40)	1	(4.05)
1	(550)	−1	(25)	−1	(5)	−1	(40)	−1	(1.28)
−1	(415)	1	(95)	−1	(5)	−1	(40)	−1	1.24
1	(550)	1	(95)	−1	(5)	−1	(40)	1	(4.05)
−1	(415)	−1	(25)	1	(15)	−1	(40)	−1	(1.28)
1	(550)	−1	(25)	1	(15)	−1	(40)	1	(4.05)
−1	(415)	1	(95)	1	(15)	−1	(40)	1	(4.05)
1	(550)	1	(95)	1	(15)	−1	(40)	−1	(1.28)
−1	(415)	−1	(25)	−1	(5)	1	(60)	−1	(1.28)
1	(550)	−1	(25)	−1	(5)	1	(60)	1	(4.05)
−1	(415)	1	(95)	−1	(5)	1	(60)	1	(4.05)
1	(550)	1	(95)	−1	(5)	1	(60)	−1	(1.28)
−1	(415)	−1	(25)	1	(15)	1	(60)	1	(4.05)
1	(550)	−1	(25)	1	(15)	1	(60)	−1	(1.28)
−1	(415)	1	(95)	1	(15)	1	(60)	−1	(1.28)
1	(550)	1	(95)	1	(15)	1	(60)	1	(4.05)
−2	(347.5)	0	(60)	0	(10)	0	(50)	0	(2.665)
2	(617.5)	0	(60)	0	(10)	0	(50)	0	(2.665)
0	(482.5)	−2	(−10)	0	(10)	0	(50)	0	(2.665)
0	(482.5)	2	(130)	0	(10)	0	(50)	0	(2.665)
0	(482.5)	0	(60)	−2	(0)	0	(50)	0	(2.665)
0	(482.5)	0	(60)	2	(20)	0	(50)	0	(2.665)
0	(482.5)	0	(60)	0	(10)	−2	(30)	0	(2.665)
0	(482.5)	0	(60)	0	(10)	2	(70)	0	(2.665)
0	(482.5)	0	(60)	0	(10)	0	(50)	−2	(−.105)
0	(482.5)	0	(60)	0	(10)	0	(50)	2	(5.435)
0	(482.5)	0	(60)	0	(10)	0	(50)	0	(2.665)

8.11 spherical ccd $\alpha = \sqrt{2} = 1.414$
  2^2 factorial portion, 4 axial runs, one center run

  design variables (natural variables)

Inlet Temp.		Reflux Ratio	
−1	(550)	−1	(4)
1	(600)	−1	(4)
−1	(550)	1	(8)
1	(600)	1	(8)
−1.414	(539.6)	0	(6)
1.414	(610.4)	0	(6)
0	(575)	−1.414	(3.2)
0	(575)	1.414	(8.8)
0	(575)	0	(6)

8.13 a spherical ccd $\alpha = \sqrt{3} = 1.732$
  2^3 factorial portion, 8 axial runs, one center run

design variables (natural variables)

Catalyst (%)	Temp. (°F)	Pressure (psig)
−1 (1.5)	−1 (250)	−1 (25)
1 (3.0)	−1 (250)	−1 (25)
−1 (1.5)	1 (280)	−1 (25)
1 (3.0)	1 (280)	−1 (25)
−1 (1.5)	−1 (250)	1 (40)
1 (3.0)	−1 (250)	1 (40)
−1 (1.5)	1 (280)	1 (40)
1 (3.0)	1 (280)	1 (40)
−1.732 (.95)	0 (265)	0 (32.5)
1.732 (3.55)	0 (265)	0 (32.5)
0 (2.25)	−1.732 (239)	0 (32.5)
0 (2.25)	1.732 (291)	0 (32.5)
0 (2.25)	0 (265)	−1.732 (19.5)
0 (2.25)	0 (265)	1.732 (45.5)
0 (2.25)	0 (265)	0 (32.5)

b rotatable ccd $\alpha = 8^{.25} = 1.682$; 2^3 factorial portion, 8 axial runs, one center run

design variables (natural variables)

Catalyst (%)	Temp. (°F)	Pressure (psig)
−1 (1.5)	−1 (250)	−1 (25)
1 (3.0)	−1 (250)	−1 (25)
−1 (1.5)	1 (280)	−1 (25)
1 (3.0)	1 (280)	−1 (25)
−1 (1.5)	−1 (250)	1 (40)
1 (3.0)	−1 (250)	1 (40)
−1 (1.5)	1 (280)	1 (40)
1 (3.0)	1 (280)	1 (40)
−1.682 (.99)	0 (265)	0 (32.5)
1.682 (3.5)	0 (265)	0 (32.5)
0 (2.25)	−1.682 (240)	0 (32.5)
0 (2.25)	1.682 (290)	0 (32.5)
0 (2.25)	0 (265)	−1.682 (20)
0 (2.25)	0 (265)	1.682 (45)
0 (2.25)	0 (265)	0 (32.5)

8.15 a Face-centered cube $\alpha = 1$

2^4 full factorial, 8 axial runs, 1 center run

Design variables (natural variables)

pH	Salt Conc.	Time	Temperature
−1 (4)	−1 (0.00)	−1 (4)	−1 (1)
1 (8)	−1 (0.00)	−1 (4)	−1 (1)
−1 (4)	1 (0.14)	−1 (4)	−1 (1)
1 (8)	1 (0.14)	−1 (4)	−1 (1)
−1 (4)	−1 (0.00)	1 (16)	−1 (1)
1 (8)	−1 (0.00)	1 (16)	−1 (1)
−1 (4)	1 (0.14)	1 (16)	−1 (1)
1 (8)	1 (0.14)	1 (16)	−1 (1)
−1 (4)	−1 (0.00)	−1 (4)	1 (2)
1 (8)	−1 (0.00)	−1 (4)	1 (2)
−1 (4)	1 (0.14)	−1 (4)	1 (2)
1 (8)	1 (0.14)	−1 (4)	1 (2)
−1 (4)	−1 (0.00)	1 (16)	1 (2)
1 (8)	−1 (0.00)	1 (16)	1 (2)
−1 (4)	1 (0.14)	1 (16)	1 (2)
1 (8)	1 (0.14)	1 (16)	1 (2)
−1 (4)	0 (0.07)	0 (10)	0 (1.5)
1 (8)	0 (0.07)	0 (10)	0 (1.5)
0 (6)	−1 (0.00)	0 (10)	0 (1.5)
0 (6)	1 (0.14)	0 (10)	0 (1.5)
0 (6)	0 (0.07)	−1 (4)	0 (1.5)
0 (6)	0 (0.07)	1 (16)	0 (1.5)
0 (6)	0 (0.07)	0 (10)	−1 (1)
0 (6)	0 (0.07)	0 (10)	1 (2)
0 (6)	0 (0.07)	0 (10)	0 (1.5)

b Rotatable ccd $\alpha = 2$

2^4 full factorial, 8 axial runs, 1 center run

Note this design uses negative salt concentrations and time (both are impossible)!

Design variables (natural variables)			
pH	Salt Conc.	Time	Temperature
−1 (4)	−1 (0.00)	−1 (4)	−1 (1)
1 (8)	−1 (0.00)	−1 (4)	−1 (1)
−1 (4)	1 (0.14)	−1 (4)	−1 (1)
1 (8)	1 (0.14)	−1 (4)	−1 (1)
−1 (4)	−1 (0.00)	1 (16)	−1 (1)
1 (8)	−1 (0.00)	1 (16)	−1 (1)
−1 (4)	1 (0.14)	1 (16)	−1 (1)
1 (8)	1 (0.14)	1 (16)	−1 (1)
−1 (4)	−1 (0.00)	−1 (4)	1 (2)
1 (8)	−1 (0.00)	−1 (4)	1 (2)
−1 (4)	1 (0.14)	−1 (4)	1 (2)
1 (8)	1 (0.14)	−1 (4)	1 (2)
−1 (4)	−1 (0.00)	1 (16)	1 (2)
1 (8)	−1 (0.00)	1 (16)	1 (2)
−1 (4)	1 (0.14)	1 (16)	1 (2)
1 (8)	1 (0.14)	1 (16)	1 (2)
−2 (2)	0 (0.07)	0 (10)	0 (1.5)
2 (10)	0 (0.07)	0 (10)	0 (1.5)
0 (6)	−2 (−0.07)	0 (10)	0 (1.5)
0 (6)	2 (0.21)	0 (10)	0 (1.5)
0 (6)	0 (0.07)	−2 (−2)	0 (1.5)
0 (6)	0 (0.07)	2 (22)	0 (1.5)
0 (6)	0 (0.07)	0 (10)	−2 (.5)
0 (6)	0 (0.07)	0 (10)	2 (2.5)
0 (6)	0 (0.07)	0 (10)	0 (1.5)

8.17 a Dependent Variable: Y = abrasion index

Indep. Variables: X_1 = hydrated silica level; X_2 = silane coupling agent level; X_3 = sulfur level

Analysis of Variance

Source	DF	Sum of Squares	Mean Square	F Value	Prob>F
Model	9	10948.93908	1216.54879	38.638	0.0001
Error	10	314.86092	31.48609		
C Total	19	11263.80000			

R-square	0.9720	overall model significant	
Adj R-sq	0.9469	R-sq 97.2% excellent	

Parameter Estimates

Variable	DF	Parameter Estimate	Standard Error	T for H0: Parameter=0	Prob > |T|
INTERCEP	1	139.119239	2.28195549	60.965	0.0001
X1	1	16.493645	1.53670069	10.733	0.0001
X2	1	17.880765	1.53670069	11.636	0.0001
X3	1	10.906538	1.53670069	7.097	0.0001
X1X1	1	−4.009601	1.54406406	−2.597	0.0266
X2X2	1	−3.447106	1.54406406	−2.232	0.0496
X3X3	1	−1.572121	1.54406406	−1.018	0.3326
X1X2	1	5.125000	1.98387538	2.583	0.0273
X1X3	1	7.125000	1.98387538	3.591	0.0049
X2X3	1	7.875000	1.98387538	3.970	0.0026

all terms in second-order model important except x3*x3.

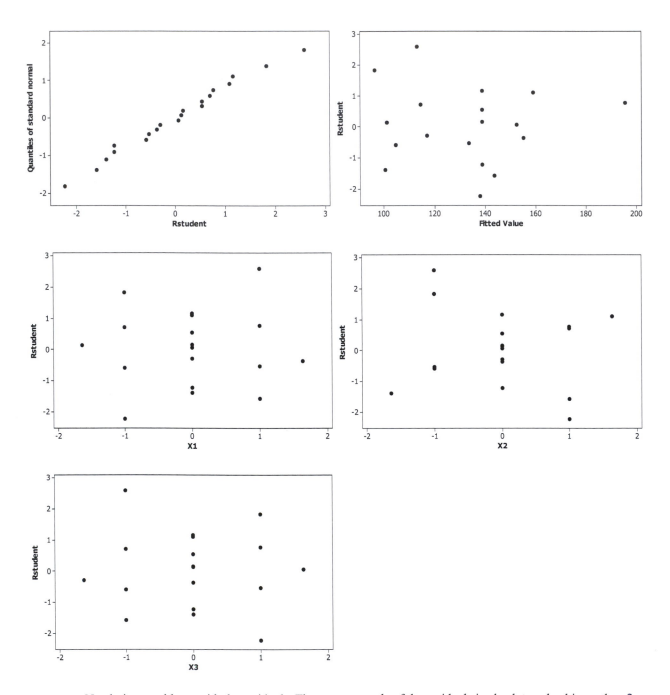

No obvious problems with the residuals. There are a couple of the residuals in absolute value bigger than 2 that deserve further investigation.

Excel Solver using the full 2nd order model with spherical-constraint.
$x1 = .949$ $x2 = 1.04$ $x3 = 1.01$ $\hat{y} = 195.56$

8.19 Dependent Variable: Y = porosity

Independent Variables: X_1 = furnace temperature; X_2 = die close time

Analysis of Variance

Source	DF	Sum of Squares	Mean Square	F Value	Prob>F
Model	5	47.36111	9.47222	33.000	0.0080
Error	3	0.86111	0.28704		
C Total	8	48.22222			

R-square	0.9821	overall model significant
Adj R-sq	0.9524	R-sq 98.21% excellent

Parameter Estimates

Variable	DF	Parameter Estimate	Standard Error	T for H0: Parameter=0	Prob > |T|
INTERCEP	1	16.888889	0.39933072	42.293	0.0001
X1	1	-2.666667	0.21872244	-12.192	0.0012
X2	1	-0.500000	0.21872244	-2.286	0.1063
X1X1	1	-0.333333	0.37883838	-0.880	0.4437
X2X2	1	1.166667	0.37883838	3.080	0.0542
X1X2	1	0.250000	0.26787919	0.933	0.4195

X1 is an important factor; X2 and X2*X2 are marginally important.

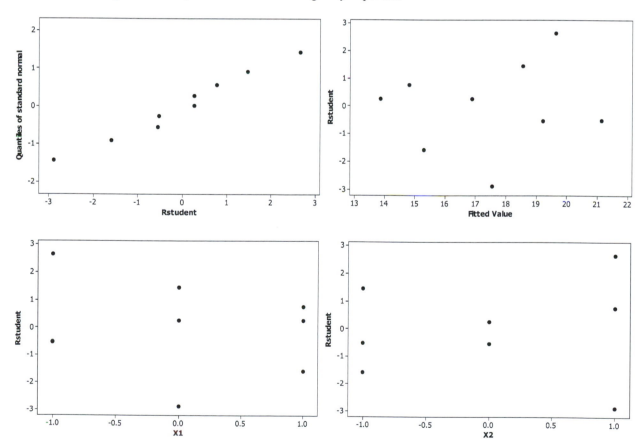

There are no clear problems form the residual plots. There are two residuals that deserve further investigation.

Refit model with x1 x2 and x2*x2.

Variable	DF	Parameter Estimate	Standard Error	T for H0: Parameter=0	Prob > |T|
INTERCEP	1	16.666667	0.29814240	55.902	0.0001
X1	1	-2.666667	0.21081851	-12.649	0.0001
X2	1	-0.500000	0.21081851	-2.372	0.0638
X2X2	1	1.166667	0.36514837	3.195	0.0241

Excel Solver

Response to be minimized at x1 = 1 and x2 = .214225, $\hat{y}$ = 13.95

8.21 The analysis was done using coded units.

Estimated Regression Coefficients for SP. Vol.

Term	Coef	SE Coef	T	P
Constant	459.996	2.724	168.856	0.000
A	10.081	1.953	5.161	0.001
B	3.231	1.953	1.654	0.132
C	9.966	1.953	5.102	0.001
D	5.182	1.953	2.653	0.026
E	0.289	1.953	0.148	0.886
F	-0.812	1.953	-0.416	0.687
A*A	-5.966	3.852	-1.549	0.156
B*B	0.034	3.852	0.009	0.993
C*C	-2.216	3.852	-0.575	0.579
D*D	-6.841	3.852	-1.776	0.109
E*E	2.659	3.852	0.690	0.507
F*F	-6.466	3.852	-1.679	0.128
A*B	-3.506	3.772	-0.929	0.377
A*C	0.303	3.772	0.080	0.938
A*D	-3.939	3.772	-1.044	0.324
A*E	-0.216	3.772	-0.057	0.956
A*F	-0.909	3.772	-0.241	0.815
B*E	0.043	3.772	0.011	0.991
B*F	0.216	3.772	0.057	0.956

S = 5.447 R-Sq = 89.1% R-Sq(adj) = 66.2%

Analysis of Variance for SP. Vol.

Source	DF	Seq SS	Adj SS	Adj MS	F	P
Regression	19	2190.86	2190.86	115.309	3.89	0.021
Linear	6	1858.42	1858.42	309.736	10.44	0.001
Square	6	272.36	272.36	45.393	1.53	0.272
Interaction	7	60.09	60.09	8.584	0.29	0.942
Residual Error	9	267.02	267.02	29.669		
Total	28	2457.88				

Important terms are A, B, C, D, F, A*A, D*D, and F*F

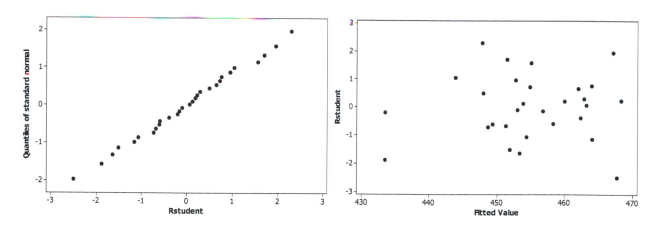

There are no clear problems form the residual plots. There are two residuals that deserve further investigation.

8.23 The analysis was done using coded units.

Estimated Regression Coefficients for ln(standard deviation)

Term	Coef	SE Coef	T	P
Constant	-1.98120	0.3180	-6.230	0.001
Temperature	0.05720	0.1220	0.469	0.656
Pressure	0.08569	0.1220	0.702	0.509
Humidity	-0.08882	0.1220	-0.728	0.494
Temperature*Temperature	-0.54362	0.1482	-3.669	0.010
Pressure*Pressure	-0.54422	0.1482	-3.673	0.010
Humidity*Humidity	0.09604	0.1482	0.648	0.541
Temperature*Pressure	0.03128	0.1595	0.196	0.851
Temperature*Humidity	0.13719	0.1595	0.860	0.423
Pressure*Humidity	0.41059	0.1595	2.575	0.042

```
S = 0.451030    PRESS = 8.24406
R-Sq = 86.44%   R-Sq(pred) = 8.38%   R-Sq(adj) = 66.09%
```

Analysis of Variance for ln(std)

Source	DF	Seq SS	Adj SS	Adj MS	F	P
Regression	9	7.7774	7.7774	0.86416	4.25	0.046
Linear	3	0.2527	0.2527	0.08423	0.41	0.749
Square	3	6.0176	6.0176	2.00588	9.86	0.010
Interaction	3	1.5071	1.5071	0.50237	2.47	0.159
Residual Error	6	1.2206	1.2206	0.20343		
Lack-of-Fit	5	0.8347	0.8347	0.16694	0.43	0.811
Pure Error	1	0.3859	0.3859	0.38586		
Total	15	8.9980				

Important terms are Temperature, Pressure, Humidity, Temperature^2, Pressure^2, and Pressure*Humidity

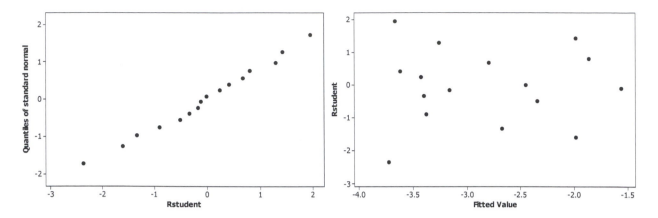

The residual plots show no departures from the assumptions.

8.25 The analysis was done using coded units.

Estimated Regression Coefficients for Velocity

Term	Coef	SE Coef	T	P
Constant	799.995	14.516	55.111	0.000
Weight	0.839	8.889	0.094	0.928
Distance	18.838	8.889	2.119	0.088
Force	71.429	8.889	8.035	0.000
Weight*Weight	9.450	13.085	0.722	0.503
Distance*Distance	11.332	13.085	0.866	0.426
Force*Force	-63.505	13.085	-4.853	0.005
Weight*Distance	0.742	12.571	0.059	0.955
Weight*Force	-4.360	12.571	-0.347	0.743
Distance*Force	-8.333	12.571	-0.663	0.537

S = 25.1425   PRESS = 45375.9
R-Sq = 95.04% R-Sq(pred) = 28.76% R-Sq(adj) = 86.11%

Analysis of Variance for Velocity

Source	DF	Seq SS	Adj SS	Adj MS	F	P
Regression	9	60532.7	60532.7	6725.9	10.64	0.009
Linear	3	43661.0	43661.0	14553.7	23.02	0.002
Square	3	16515.7	16515.7	5505.2	8.71	0.020
Interaction	3	356.0	356.0	118.7	0.19	0.900
Residual Error	5	3160.7	3160.7	632.1		
Lack-of-Fit	3	2782.9	2782.9	927.6	4.91	0.174
Pure Error	2	377.9	377.9	188.9		
Total	14	63693.4				

Important terms are Distance, Force, and Force*Force

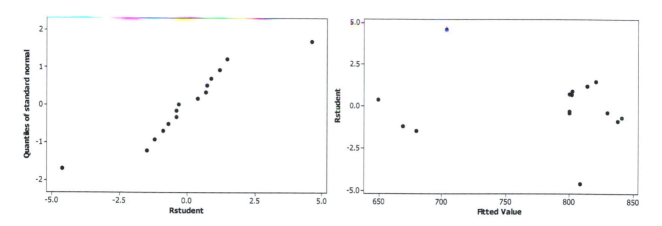

There are no clear problems form the residual plots. There are two residuals that deserve further investigation.

8.27 The analysis was done using coded units.

Estimated Regression Coefficients for Convexity

Term	Coef	SE Coef	T	P
Constant	0.35978	0.16186	2.223	0.035
Arc Length	-1.23262	0.09912	-12.436	0.000
CTWD	-0.04608	0.09912	-0.465	0.646
Angle	-0.32850	0.09912	-3.314	0.003
Diameter	-0.27730	0.09912	-2.798	0.010
Speed	0.85922	0.09912	8.669	0.000
Arc Length*Arc Length	-0.23172	0.13420	-1.727	0.097
CTWD*CTWD	-0.07986	0.13420	-0.595	0.557
Angle*Angle	0.02454	0.13420	0.183	0.856
Diameter*Diameter	-0.34001	0.13420	-2.534	0.018
Speed*Speed	-0.18882	0.13420	-1.407	0.172
Arc Length*CTWD	0.31800	0.19823	1.604	0.121
Arc Length*Angle	-0.31337	0.19823	-1.581	0.126
Arc Length*Diameter	-0.13885	0.19823	-0.700	0.490
Arc Length*Speed	0.48155	0.19823	2.429	0.023
CTWD*Angle	-0.06690	0.19823	-0.337	0.739
CTWD*Diameter	0.13698	0.19823	0.691	0.496
CTWD*Speed	-0.02598	0.19823	-0.131	0.897
Angle*Diameter	-0.39453	0.19823	-1.990	0.058
Angle*Speed	0.22620	0.19823	1.141	0.265
Diameter*Speed	0.18680	0.19823	0.942	0.355

```
S = 0.396463    PRESS = 15.5491
R-Sq = 91.71%   R-Sq(pred) = 67.20%   R-Sq(adj) = 85.08%
```

Analysis of Variance for Convexity

Source	DF	Seq SS	Adj SS	Adj MS	F	P
Regression	20	43.4784	43.4784	2.17392	13.83	0.000
Linear	5	39.1126	39.1126	7.82252	49.77	0.000
Square	5	1.5013	1.5013	0.30026	1.91	0.128
Interaction	10	2.8645	2.8645	0.28645	1.82	0.108
Residual Error	25	3.9296	3.9296	0.15718		
Lack-of-Fit	20	3.8635	3.8635	0.19317	14.62	0.004
Pure Error	5	0.0661	0.0661	0.01322		
Total	45	47.4079				

Important terms are Arc Length, Angle, Diameter, Speed, Arc Length*Arc Length, Diameter*Diameter, Arc Length*Speed, and Angle*Diameter.

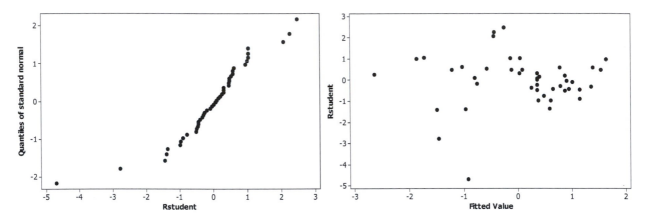

The residual plots show no departures from the assumptions. However, there are several outliers.

8.29 The analysis was done using coded units.

Estimated Regression Coefficients for Yield

Term	Coef	SE Coef	T	P
Constant	85.5380	2.670	32.035	0.000
Temp.	-6.5583	1.724	-3.805	0.002
CHPTAC	0.8542	1.724	0.496	0.628
STP	-0.1300	1.724	-0.075	0.941
Time	-1.0875	1.724	-0.631	0.538
Temp.*Temp.	0.9181	2.344	0.392	0.701
CHPTAC*CHPTAC	2.8868	2.344	1.231	0.238
STP*STP	-4.3394	2.344	-1.851	0.085
Time*Time	0.1493	2.344	0.064	0.950
Temp.*CHPTAC	2.9250	2.985	0.980	0.344
Temp.*STP	-3.4950	2.985	-1.171	0.261
Temp.*Time	-4.4650	2.985	-1.496	0.157
CHPTAC*STP	-0.0600	2.985	-0.020	0.984
CHPTAC*Time	-0.2325	2.985	-0.078	0.939
STP*Time	0.5000	2.985	0.167	0.869

S = 5.97069   PRESS = 2255.37
R-Sq = 65.01%   R-Sq(pred) = 0.00%   R-Sq(adj) = 30.03%

Analysis of Variance for Yield

Source	DF	Seq SS	Adj SS	Adj MS	F	P
Regression	14	927.4	927.4	66.25	1.86	0.129
Linear	4	539.3	539.3	134.82	3.78	0.027
Square	4	224.1	224.1	56.02	1.57	0.236
Interaction	6	164.1	164.1	27.34	0.77	0.608
Residual Error	14	499.1	499.1	35.65		
Lack-of-Fit	10	351.5	351.5	35.15	0.95	0.570
Pure Error	4	147.6	147.6	36.89		
Total	28	1426.5				

Important terms are Temperature, STP, and STP*STP.

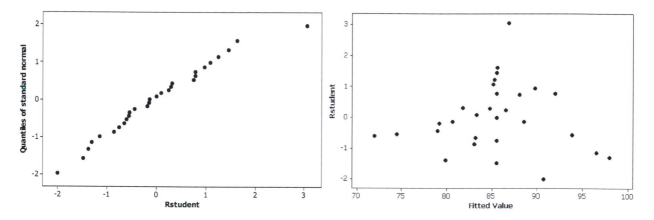

The residual plots show no departures from the assumptions. However, there is one outlier.

8.31 a Solution obtained from Design Expert software

$$\hat{y}_1 = 6.78 - 1.17x_1 + .05x_2 - .17x_1^2 - 4.17x_2^2$$
$$\hat{y}_2 = 17.44 - 2.67x_1 - .50x_2$$
$$\hat{y}_3 = 94.44 + 10.50x_1 + 3x_2$$

$$d_1 = \begin{cases} 1 & \hat{y}_1 = 0 \\ 0 & \text{otherwise} \end{cases}$$

$$d_2 = \begin{cases} 1 & \hat{y}_2 < 0 \\ (21 - \hat{y}_2)/(21 - 0) & 0 \le \hat{y}_2 \le 21 \\ 0 & \hat{y}_2 > 21 \end{cases}$$

$$d_3 = \begin{cases} 1 & \hat{y}_3 < 0 \\ (108 - \hat{y}_3)/(108 - 0) & 400 \le \hat{y}_3 \le 500 \\ 0 & \hat{y}_3 > 108 \end{cases}$$

Possible settings		Predicted results		
x_1	x_2	$\hat{y}_1$	$\hat{y}_2$	$\hat{y}_3$
.4	−1	1.61723	16.8767	95.6489
−.07	1	3.19435	17.3670	96.6828

b Solution obtained from the Excel Solver
Objectives: $\hat{y}_1 = 0, \hat{y}_2 \leq 21, \hat{y}_3 \leq 108$
Constraints: subject to x_1, x_2 between -1 and 1.

Recommended settings		Predicted results		
x_1	x_2	$\hat{y}_1$	$\hat{y}_2$	$\hat{y}_3$
1	-1	.77	15.27	101.94

8.33 a Solution obtained from Design Expert software, using Natural Variables
$\hat{y}_1 = 5541.23 - 28.98x_1 - 17.29x_2 + 7.1x_3 + .026x^2_1 - .0021x^2_2 - .069x^2_3 + .072x_1x_2 - .058x_1x_3 + .049x_2x_3$
$\hat{y}_2 = 1776000 - 8985.66x_1 - 6009.01x_2 + 2277.27x_3 + 13.72\,x^2_1 + 6.06x^2_2 - 22.64x^2_3 - 110.71x_1x_2 - 3.57x_1 + 4.81x_2x_3$

$$d_1 = \begin{cases} 0 & \hat{y}_1 < 94 \\ (\hat{y}_1 - 94)/(100 - 94) & 94 \leq \hat{y}_1 \leq 100 \\ 1 & \hat{y}_1 > 100 \end{cases}$$

$$d_2 = \begin{cases} 0 & \hat{y}_2 < 4000 \\ (\hat{y}_2 - 4000)/(5700 - 4000) & 4000 \leq \hat{y}_2 \leq 5700 \\ 1 & \hat{y}_2 > 5700 \end{cases}$$

Recommended settings			Predicted results	
x_1	x_2	x_3	$\hat{y}_1$	$\hat{y}_2$
220.67	299.42	69.25	104.454	6813.74
224.24	299.47	59.23	104.409	7147.30

b Solution obtained from the Excel Solver
Objectives: maximize $\hat{y}_1, \hat{y}_2 \geq 4000$

Recommended settings		
x_1	x_2	x_3
230	300	62.82

8.35 a Solution is
$\hat{y}_1 = -1.83 + .057x_1 + .086\,x_2 - .089\,x_3 - .584\,x_1x_1 - .585\,x_2x_2 + .411x_2x_3$
$\hat{y}_2 = 9.98 + .051\,x_1 + .048\,x_2 + .048\,x_3$

where $\hat{y}_1$ is the natural log of the standard deviation and $\hat{y}_2$ is volume

Recommended Settings			Predicted results	
x_1	x_2	x_3	y_1	y_2
1.688	-1.688	.298	-5.42	10

8.37

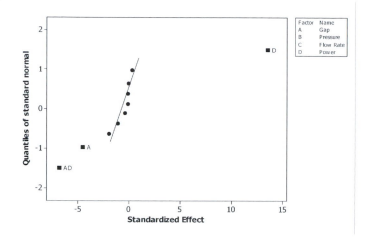

Dependent Variable: Y = etch rate
Indep. Vars.: A = gap, spacing between the anode and the cathode; D = power applied to the cathode

Analysis of Variance

Source	DF	Sum of Squares	Mean Square	F Value	Prob>F
Model	3	510563.18750	170187.72917	97.913	0.0001
Error	12	20857.75000	1738.14583		
C Total	15	531420.93750			

R-square	0.9608	Overall model significant
Adj R-sq	0.9509	R-sq 96.08% excellent

Parameter Estimates

Variable	DF	Parameter Estimate	Standard Error	T for H0: Parameter=0	Prob > |T|
INTERCEP	1	776.062500	10.42276905	74.458	0.0001
A	1	-50.812500	10.42276905	-4.875	0.0004
D	1	153.062500	10.42276905	14.685	0.0001
AD	1	-76.812500	10.42276905	-7.370	0.0001

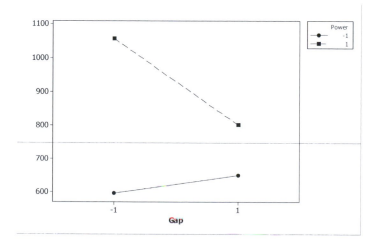

The interaction plot suggests setting the level of factor A (Gap) at its high level 1.20cm to reduce the effect of the noise factor D (Power). A compromise may need to be reached since factor A has a negative main effect.

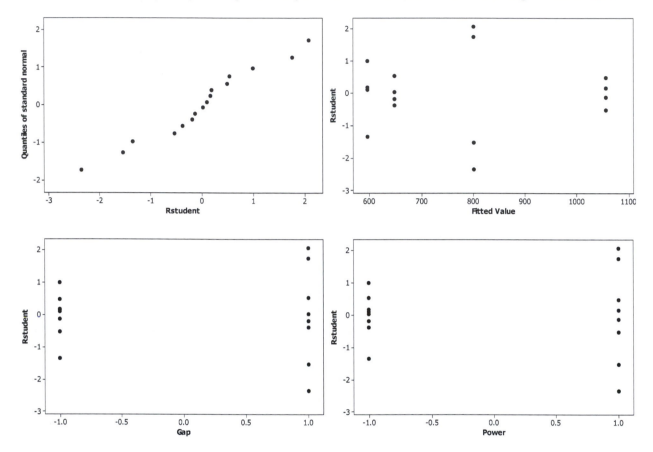

There is an Indication of an increase in the variance with A and D. The normal probability plot follows a straight line.

8.39 Dependent Variable: Y = distance
Independent Variables: X_1 = hook; X_2 = arm length; X_3 = start angle; X_4 = stop angle

Analysis of Variance

Source	DF	Sum of Squares	Mean Square	F Value	Prob>F
Model	7	71718.32000	10245.47429	11258.763	0.0001
Error	16	14.56000	0.91000		
C Total	23	71732.88000			

Root MSE	0.95394	R-square	0.9998	
Dep Mean	78.80000	Adj R-sq	0.9997	
C.V.	1.21058			

Parameter Estimates

Variable	DF	Parameter Estimate	Standard Error	T for H0: Parameter=0	Prob > |T|
INTERCEP	1	78.800000	0.19472202	404.679	0.0001
X1	1	42.200000	0.19472202	216.719	0.0001
X2	1	10.191667	0.19472202	52.340	0.0001
X3	1	26.600000	0.19472202	136.605	0.0001
X4	1	5.225000	0.19472202	26.833	0.0001
X1X2	1	9.858333	0.19472202	50.628	0.0001
X1X3	1	14.050000	0.19472202	72.154	0.0001
X1X4	1	8.608333	0.19472202	44.208	0.0001

The residual analysis indicated no major problems.

Dependent Variable: S = standard deviation
Independent Variables: X_1 = hook; X_2 = arm length; X_3 = start angle; X_4 = stop angle

Analysis of Variance

Source	DF	Sum of Squares	Mean Square	F Value	Prob>F
Model	4	0.90790	0.22698	5.286	0.1013
Error	3	0.12881	0.04294		
C Total	7	1.03671			

Root MSE	0.20721	R-square	0.8757	
Dep Mean	0.88341	Adj R-sq	0.7101	
C.V.	23.45621			

Parameter Estimates

Variable	DF	Parameter Estimate	Standard Error	T for H0: Parameter=0	Prob > |T|
INTERCEP	1	0.883409	0.07326129	12.058	0.0012
X1	1	-0.198539	0.07326129	-2.710	0.0732
X2	1	-0.226604	0.07326129	-3.093	0.0536
X3	1	0.092147	0.07326129	1.258	0.2975
X4	1	0.119287	0.07326129	1.628	0.2020

Residual analysis shows a couple of studentized residuals near −3.

Use Excel Solver
Objectives: $\hat{y}$ = 100, minimize s

$x_1 = 1$ $x_2 = 1$ $x_3 = -.66954$ $x_4 = -1$ result: s = .277283

Made in the USA
Lexington, KY
20 March 2012